Neurocultural Health and Wellbeing

Series Editors

Lorenzo Lorusso, Neurology Unit
A.S.S.T. Lecco-Merate
Merate, Italy

Bruno Colombo, DIMER, Neurologia
Ospedale San Raffaele
Milano, Italy

Alessandro Porro, Dip. di Scienze Cliniche e di Comunità
University of Milan
Milano, Italy

Nicholas Wade, Department of Psychology
University of Dundee
Dundee, UK

Aim of this devoted book Series in neurology is to highlight the relationship between neuroscience and culture. Nowadays, there is more evidence of how our brain is influenced by the various artistic and cultural disciplines, both in the form of entertainment as well as – and above all – through the emotions and gratification which lead to wellbeing. The emotional mechanisms, the different cultural manifestations provided, are due to the activation of a perceptual and cognitive range that constitute the basis of the social behaviors. All this makes us aware of the benefits the arts have on personal and collective health.

This concept was already known to philosophers in ancient times, who were convinced that inner balance was influenced by culture and in particular by music.

During the centuries, men understood that the *cultivation of the spirit*, or *humanitas*, had a certain role on behavior and from the beginning of modern experimental science, in the fifteenth, the spread of the notions of neuroanatomy allowed artists to get closer to the knowledge of the brain's mechanisms and reveal the emotional and empathic responses at the basis of creativity, and indeed of their own psychophysical wellbeing. A dialogue between the science of the mind and artistic disciplines was born. The results of this meeting made it possible to better define which mental processes are involved when we come into contact with the various humanistic disciplines, and how they can be applied, for instance, to the treatment of mental disorders and neuropsychiatric diseases.

The goal of this Series is both to prove the role of biological brain mechanisms and the influence of various artistic forms on clinical practice, especially in neuro-psychiatric disorders, as well as to trace different therapeutic and psycho-physical well-being applications based on scientific evidence from medical literature.

Volumes of the Series will be edited by experts under the supervision of an international editorial committee. Each book, focused on a specific discipline, will provide knowledge on relationship between the brain activity and different forms of language, communication, art. This close inter-relationship with specific focus on different forms of art will explain the effectiveness of this kind of approaches in neuro-psychiatric diseases.

This Series will allow to understand how the culture is one of the fundamental tools to improve general well-being, quality of life and motivation in neurological diseases.

This Series also find a correlation with SDG3 goal "ensuring healthy lives and promoting wellbeing for all at all ages" for the progress of the health and wellbeing considering that neurological diseases are on the rise worldwide including in the developing countries.

Bruno Colombo
Editor

The Musical Neurons

 Springer

Editor
Bruno Colombo
Department of Neurology, Headache Clinical and Research Center
Ospedale San Raffaele, Vita-Salute University
Milano, Italy

ISSN 2731-4464 ISSN 2731-4472 (electronic)
Neurocultural Health and Wellbeing
ISBN 978-3-031-08134-7 ISBN 978-3-031-08132-3 (eBook)
https://doi.org/10.1007/978-3-031-08132-3

This Springer imprint is published by the registered company Springer Nature Switzerland AG
The registered company address is: Gewerbestrasse 11, 6330 Cham, Switzerland

*To my wife Paola, who has enabled me to
succeed in achieving my lifelong goals.
To my sons, Lorenzo, Francesco e Stefano.
Your constant love inspires me continually.
In loving memory of my parents, without
whose support I might never had made this
journey.
To my patients, who have taught me so much
about dignity, resilience and courage.*

Preface

Inventas vitam iuvat excoluisse per artes
Let us improve life through science and art. (Publius Vergilius Maro (70 AC–19 AC))

I would like for this book to be a text not solely for physicians or specialised neurologists and also not an insight merely for experts and musicologists.

Rather I would wish this to be read by minds curious to explore the interweaving that music can create in our mental sphere and by extension how "organised sounds" can be used for our well-being and people's health.

In ceremonies, religion and communication percussive rhythm has always had significance because it has forever recalled the sounds of our body such as the beat of the heart and the sound of steps.

Music, immaterial and aerial, represents the sound of time, the soul, the absolute.

Aristotle himself argued that music was able to influence the character of the soul, producing or crippling acts of will, and creating inebriation and ecstasy.

Discovering that a musical composition can generate, build and consolidate neuronal structures can't surprise us.

If anything, our brain's musicality does nothing but reiterate how vital it is to nourish it.

We must see music as substance, nutrient and, most of all, as factor to stimulate growth.

A growth not only of synapses but above all of emotivity and co-participation.

"let's start" courtesy of Lorenzo Colombo

Milano, Italy   Bruno Colombo

Introduction

Music for well-being: feed the brain!

We live in a world where music is constantly present. We listen to music not only to relax, but also to move and to dance. Music is powerful when emotionally highlighting a scene in a movie and accompanies us on the radio when we are travelling by car. Our national anthems are music as well as the hymns of our religious ceremonies. Music can be entertainment, communication and a commentary to happy or sad events in or lives. Undeniably, music provides and provokes a response, leading to marked changes in our movements and emotions.

In the last decades, research in the fields of neurophysiology, neurorehabilitation and neuroimaging has investigated what music does to our brain and how it can help its activity. Notably, it has been demonstrated how music can be beneficial to physical well-being and to keeping a positive attitude to life.

Unquestionably, music, listened to or actively practised, is able to prompt and stimulate movement and to give rhythm to dance. In doing so, it activates both the motor cortex and the basal ganglia, delivering joy and energy. Music helps to create a sense of belonging and has a social and cohesive purpose (e.g. think of people singing in a choir). It promotes communication and allows to express emotions. Music can provoke clear sensory stimuli, helpful, for example, to the visually impaired. For all these reasons, music undoubtedly has a therapeutic value and, depending on its cultural role, facilitates social learning, human interaction and cohesion within society, and emotional well-being (Box 1).

Most importantly, music favours neuroplasticity: morphometric MRI studies in musicians with early commencement of musical training (<7 years of age) compared both to musicians who started later and control groups reveals a significantly larger anterior corpus callosum size, confirming an increased interhemispheric communication [1]. In other studies, a pattern of differences in the brain grey matter distribution between professional musicians, amateurs and non-musicians was observed. In particular, an increase in grey matter volume was detected in motor-related auditory and visual regions, suggesting a possible direct use-dependent relationship between long-term musical training and related structural changes in specific brain areas [2]. Furthermore, it was demonstrated that outstanding musical ability is associated with increased leftward asymmetry of cortex (planum temporale). This area is involved in music perception and results of this study lends anatomical support of a difference in lateralisation of musical processing between musicians and non-musicians, with more left-lateralised representation in musicians [3].

But music is also able to help in curing neurological diseases. Music therapy is defined as using "clinical and evidence-based music interventions to accomplish individualized goals within a therapeutic relationship by a credentialed professional who has completed an approved therapy program. Music therapy is a reflexive process wherein the therapist helps the patient to optimize the patient's health, using various facets of music experience and the relationships formed through them as the impetus for change [4]". Such musical experiences may consist of listening to music, playing or improvising music on an instrument, composing music and using music in association with other means such as dance or images.

Music therapy can significantly improve anxiety during treatment. Rhythmic entrainment of motor function can enhance recovery of movements in patients affected by Parkinson's disease or stroke patients. In particular, music therapy could be a useful treatment to help improve hand function and daily life activities in stroke patients, especially within 6 months after a neurological event [5]. In Parkinson's disease, music therapy has shown benefits on gait automation, motor performance, dynamic balance, walking speed and an increased pitch length [6]. Music therapy is also a possible action for cognition in Alzheimer's disease: many studies have proved that music therapy can reduce cognitive decline, especially in autobiographical and episodic memories, psychomotor speed, executive function domains and global cognition [7]. Furthermore, music therapy improves sleep quality, particularly the duration of sleep and sleep efficiency, both in adults and young people [8].

For all these reasons, it is important to learn and enjoy music: music is not a sort of non-essential extra-discipline and it is important for the functioning of our culture and society. Music elicits sentiments and is tied to emotions, giving rise to a variety of sensations such as nostalgic remembering or pleasurable chills. Music is the language of feelings, revealing "the nature of feelings with a detail and truth that language cannot approach". Furthermore, participatory music engagement supports mental well-being through managing and expressing emotions, facilitating self-development, providing respite and promoting connections.

We possess musical neurons, it is our responsibility to feed them.

Box 1. Music for Sharing: Composers and Society
Until the mid-eighteenth century, people could listen to sacred vocal music in church during religious services. Otherwise, composed music was often for those who could afford it, for the court of kings and emperors or for those who visited the opera, which since the beginning of the seventeenth century, was a vocal form of drama. Artistic productions were mostly commissioned by members of the aristocracy, and the contents were inspired by norms and ideals of beauty, perfection, distinction, and elegance that were the formal canon the aristocracy, as the ruling social class, aspired to. Instrumental music was limited to compositions for teaching purposes or short intermezzos of operas. When, around 1750 and following the transformations in the wake of the

French Revolution, the first concert halls were built and the demand for compositions that could reach a wider public increased. Musicians are now able to share their music with more people and write music to perform in appropriate meeting locations with better listening conditions. Music is hence composed with a structure that is understandable, accessible and enjoyable. The new "commissioner" is now the paying public, maybe less demanding but larger. Furthermore, the composer feels free to create and present his style and, in this new society that accepts a multiplicity of trends, the composer takes the responsibility of a spiritual and moral guide, a social lighthouse, a critical and autonomous entity. The result of these changes is the so-called sonata form where music can be structured with specific characteristics and that we then find in symphonies, sonatas for solo instruments and in string quartets. The pattern, in all its variations, presents a musical theme followed by its development and its reprise. Composers can employ structural and temporal aspects of music attempting to attain the listener's response. The final goal is to create expectations and anticipation, affecting an individual's emotional response to the music. The audience recognises these concatenations and participates expecting an evolution. It is a sort of dialogue between the composer, the performer and the listener. However, in concert halls this is a shared experience, a path followed simultaneously in search of a pleasurable and unique moment.

Photo: "The stage" courtesy of Lorenzo Colombo

References

1. Schlaug G. The brain of musicians: a model for functional and structural adaptation. Ann NY Acad Sci 2001;930:281–99
2. Gaser C, Schlaug G. Brain structures differ between musicians and non-musicians. J Neurosci 2003;23(27):9240–5
3. Ohnishi T, Matsuda H, Asadat T, et al. Functional anatomy of musical perception in musicians. Cereb Cortex 2001;11(8):754–60
4. American Music therapy Association. Definition of music therapy. 2011. http://musictherapy.org
5. Altenmuller E, Schneider S, Marco-Pallares PW, Munte TF. Neural reorganization underlies improvement in stroke induced motor dysfunction by music supported therapy. Ann NY Acad Sci 2009;1169:395-405
6. Dalla Bella SD, Benoit CE, Farrugian C et al. Effects of musically cued gait training in Parkinson's disease: beyond a motor benefit. Ann NY Acad Sci 2015;1337:77–85
7. Colombo B. Is there an artistic treatment for neurological diseases? The paradigm of music therapy. In: Colombo B. editor. Brain and art; 2019. Berlin: Springer International. doi: https://doi.org/10.1007/978-3-030
8. Feng F, Zhang Y. Can music improve sleep quality in adults with primary insomnia? A systematic review and network meta-analysis. Int J Nurs Stud 2018;77:189–96

Department of Neurology, Headache   Bruno Colombo
Clinical and Research Center
Vita-Salute University
Milano, Italy

Acknowledgements

During the months I was preparing and writing this book, a special person gave me practical help and advice. My thanks must go next to Dr. Francesca Bernecker for the substantial assistance and encouragement. Her generosity greatly speeded the completion of this work. I offer her all my hearty thanks.

I also wish to thank the many contributors to this book for their hard work, as well as Andrea Ridolfi, editor at Springer, for his help and guidance.

Contents

About the Editor

Bruno Colombo is a board certified Neurologist with a passionate interest in headache management and patient care. He is founder both of the Italian Neurological Association for the Research on Headache (ANIRCEF) and the Italian Foundation for Headache (FICEF), and currently serves on the board of Directors.

He has authored and co-authored 200 articles on peer-reviewed journals. He has also authored three books.

He has served as Clinical Assistant and Research Worker at the National Hospital for Nervous Diseases and Institute of Neurology, London.

He is currently Coordinator and Responsible of the Headache Center at Vita-Salute University, San Raffaele Hospital in Milan.

He is academic and teacher at Vita-Salute University and lecturer at Master on Headache at Biomedical Campus in Rome.

As a musician, he was first attracted to classical music and progressive rock.

At his late teens, Dr. Colombo picked up an interest in jazz music.

By the time he began attending Civic School of Music in Milan expanding his listening and playing scope into various jazz forms.

Accomplished on guitar and bass, was also adept in drums and percussions.

During this time, he played around locally semipro with avant-garde jazz combos.

Attracted to contemporary music in adulthood, he became interested especially in the theories of composer John Cage.

At the same time, he presented programmes for several radio stations considering a broad range of topics: jazz, rock, popular and contemporary music.

During his time at the University of Milan, he first began music criticism, writing record reviews and cultural commentaries for music magazines.

He is actually involved both in promoting the study of popular music with an historical perspective, and in studying neural correlates of consonance and dissonance.

Music for Pleasure

1

Bruno Colombo

1.1   Some Narrative Reflections

What is music? And more importantly, what cultural value does music have?

Why are human beings musical?

And again, does music have a moral significance?

The definition of music in the Encyclopedia Britannica can serve us as a start: "… art concerned with combining vocal or instrumental sounds for beauty or form of emotional expression, usually according to cultural standards of rhythm, melody, and in most Western music, harmony" [1].

We see how with time the meaning of music has conceptually evolved comparing the above definition with Rousseau's in his 1767 *Dictionnaire de musique*: "… art of combining sound in a way pleasant to the ear" [2].

Rousseau's "sounds" become "vocal or musical sounds" and their combination is not anymore only pleasant to the ear but also an expression of "beauty" and of emotional activity related to perceptions that rely on the culture of the musician and the listener. There is an opening to timbre, rhythms, and sound juxtapositions, not necessarily homogeneous or defined by absolute "pleasantness" if not linked to the cultural context they are created in.

This modernization in the conception of music is due to the anthropological revolution started in the twentieth century, originating from constant and fruitful ethnomusical research.

World music scholars confirm that in all societies, no matter how small or isolated, symbolic productions carried out through sound can be found. Thus, resorting to musical expression is universal to the humankind, as language is. Just as different languages exist, there are very numerous ways of arranging sounds and musical

B. Colombo (✉)
Vita-Salute University, San Raffaele Hospital, Milano, Italy
e-mail: colombo.bruno@hsr.it

© The Author(s), under exclusive license to Springer Nature
Switzerland AG 2022
B. Colombo (ed.), *The Musical Neurons*, Neurocultural Health and Wellbeing,
https://doi.org/10.1007/978-3-031-08132-3_1

1

creations that are directly linked to their own cultural roots and the societies they have developed, layered, and evolved in.

However, music is an art with fluid boundaries. Let's think of the very idea of art: a human activity conducted and designed to create symbols or an activity aimed at enhancing an aesthetic purpose, to this effect dependent on judgments on its value, the taste of its time, or cultural uniformity?

Undoubtedly, shifts in art scenery in general are the result of the evolution of time. Throughout art history, renewal is continuous and the need to always create something new and original is relentless. Look at the progression we perceive when admiring Leonardo's *Mona Lisa*, Picasso's *Les demoiselles d'Avignon*, and Jackson Pollock's artwork. The evident changes are radical: the move from a figurative representation to a more expressive one, the freedom from conventions of perspective or verism, the use of vivid and contrasting colors, the artist's gestural and almost shamanic dimension in the making of the artwork. Nevertheless, all these works of art are readable. Also in Mondrian or Leger, we are able to recognize known geometrical shapes and colors, and in Pollock we can visualize his gesture and movement, the dramatic action, and corporality.

Applying these reflections to music, it is not difficult to see similar contrasts between a "canonically classical" piece of music as, for example, Handel's *Water Music,* and Debussy's piano *Preludes*, Stravinsky's *Sacre du Printemps*, or Xeneakis' *Pleiades*.

Debussy initiated the first big change in the twentieth century. He creates a language where musical writing is fragmented and sonority becomes a cornerstone parameter as important as harmony and rhythm. "Conceive otherwise" is how Debussy thinks of his compositions. The initial phrase of *Prelude à l'apres midi d'un faune* of 1894 is expressed without harmonic support, with a small tonal interval (triton) that then leans on seventh chords. The structures appear veiled, like arabesques, and correspondences between notes are revealed as open, not encumbered by any preconstructed direction.

Stravinsky composed his ballet *Sacre du Printemps* in the winter at the turn of 1912 and 1913, based on a vision of a sacrificial rite. A group of old wise men were reunited in a circle and watched a young girl dancing to exhaustion and death, a sacrifice to ingratiate the god of Spring. The first performance in Paris in May 1913 caused a great scandal and fierce protests from the audience that rebelled against such a new and violent music. But that score, with a duration of 35 min, emblematically marked the avant-garde of those years. It turned upside down all hitherto regarded canons of taste and beauty. The primitivity, the simplification of melodic lines in connection with the complexity of rhythmic structures, the obsessive, almost percussive orchestration, and finally the overlapping of consonances and dissonances make this ballet the obligatory step between figurative and expressionist in music. For Stravinsky, modernity is in the technique of composition, in the form itself, in the symbolism that recalls archaic and earthbound values.

In the field of music, the beginning of the twentieth century has hence led to the search for new harmonic connections, different timbres, and increasingly bold constructions able to carry out a sometimes radical opposition to established canons. In this context, atonality develops, a definitive contraposition to the major–minor tonal system and an expression of the cultural crisis of contemporary society.

Following this, only some years later, Stravinsky was accused of belonging to the past. In his *Philosophy of Modern Music*, Theodor Adorno marks him as representative of conservatism (Stravinsky and Restoration). At the opposite pole, Adorno places Arnold Schoenberg, the creator of the 12-tone technique, whom he describes as progressive (Schoenberg and Progress) [3] (Box 1.1).

In the last years of his life and his career, Stravinsky himself moved closer to this music with some complex works written in a 12-tone musical key, such as the ballet *Agon*. He thus embraces the new avant-garde, which will then move still further, as we can see from the elements and substance of contemporary music of the twentieth century.

Iannis Xenakis, for instance, forges a musical poetics seeking an "objective" noise of the world, a musical pattern as similar as possible to the sound of nature [4]. For Xenakis, "music has to be taken out of the atrophied glasshouses of tradition and brought back to nature." His compositions are often for percussion instruments, thus supporting the need for strong sound and rhythm. Listening to them means interpreting a notation based on the formal language of mathematical formulas of complex theories of randomness such as stochastic process, fluid mechanics, and statistical laws.

Still, in all of this music, we always hear contrasts, crescendos, ruptures, inversions, developments, progresses, and conclusions, which enable us to find a listening orientation in a composition, even when they refer to a language we are not always familiar with.

Photo: "Switching on Xenakis," courtesy of Lorenzo Colombo

John Cage corroborates a concept of music as aleatoric geometry (and how not to think of Mondrian's painting here): "If this word, music, is sacred and reserved for eighteenth- and nineteenth-century instruments, we can substitute a more meaningful term: organization of sound" [5]. But is it sound? In August 1952, the score of 4'33'' is first performed. The pianist sits at the piano, he clocks the times (30'', 2'23'' 1'40''), opens and closes the lid of the keyboard in correspondence, never touching a key. The score indicates "tacet." But is it music or silence? Or maybe silence doesn't exist: in those 4'33'' the listeners can maybe hear their hearts beating, a chair creaking, noises from the surroundings. The silence, which John Cage defines as "true, positive, active, affirmative… we were there before but we just weren't aware" [6]. For Cage, sounds must be just what they are and to listen means to open oneself to the totality and complexity of that which exists, without any kind of intent. Silence potentially contains all sounds. It is open to any possibility, up to the edge of uncertainty. Chance and the listener are in charge of filling the void and, thanks to silence, the multiplicity of noises find their way into the music. Is then 4'33'' a musical work or the art of contemplating time, life, and unpredictability?

However, all this change is the result of the knowledge and study of non-western cultures. As early as 1885, Ellis was pointing out that musical scales can be very diverse in different cultures and concluded that "The musical scale is not one, not "natural," not even founded necessarily on the laws of the constitution of musical sound so beautifully worked out by Helmholtz…" [7] (Box 1.2).

Photo: "A prepared piano," courtesy of Alexander Banck-Petersen

The impact this anthropological engagement had on the dominant musical culture was disrupting: western musical language is in crises. The introduction of atonality, of 12-tone scale and of a way of composing able to overcome formally constructed shapes and structures expressed the acceptance of a new concept that shows what music can be: a form of expression and of language that can be found in any culture and based on specific means of organizing sound. A form of expression is not required to necessarily follow canons of beauty or pleasantness but is expected to be coherently appropriate to its mission and purpose. In this regard, the final aesthetic is only one, but not the only one, of the critical qualities of an artistic process as symbolic as music.

Music can effectively be described as the whole of the forms and practices that a society or culture considers eligible to create an organized sound, according to its own criteria. In some African cultures, for example, it is at times difficult to distinguish between music and dance, and this can be reflected in the lack of a specific terminology. The word "kwina" used by the Kamba in Kenia, for instance, means to sing as well as to dance. On the other hand, the Somali people have three different words: "heesid" for singing songs, "ciyaarid" for sound accompanying dance, and "gabyid" for singing poetry. In the Inuit language, the word "nipi" means music, but also noise and even the sound of a speaking voice. The distinction between sound and noise made by the Basonge of Central Africa is also interesting: "when you are merry you sing, when you are angry you make noise … singing is peaceful, noise isn't" [8]. In ethnomusical and cultural terms, all this leads to an interpretation and to the conferring of musicality to sound. In the intentional act of producing a desired sound lies its musicality. This means having an organized or extrapolated plan related to a musical action with specific and linked functions such as having techniques in the phonic use of voice, having objects designed for producing organized sound, assigning a defined temporality to the sound produced. Notably, in this last definition, we can find a feature peculiar to music: music is intangible, immaterial, and incorporeal but can become concrete precisely and exclusively thanks to its specific temporality, made of rhythms, lengths, and silences inserted in the "present time," becoming however a "virtual time." As Levi-Strauss said, "music is a machine to abolish time" [9].

Box 1.1

Traditional references to tonality are revolutionized by Arnold Schoenberg in 1909 with his absolutely atonal composition *Three Piano Pieces, Op. 11*. "I aspire to a complete liberation from all symbols of coherence and logic" he said, "no more harmony seen as an architectural building block … harmony is an expression and nothing else" [10]. Later, in 1929, Schoenberg invented and applied the so-called "serial" technique to various compositions for piano (Opp. 23, 24, and 25). The 12-tone structure is based on a series of sounds, mostly 12, linked to one another according to a definite order, with the chromatic scale as its starting point. The principle this system is based on is that the same note must not reoccur before all other 12 notes have been played. The harmonic rules' tonality had accustomed us to have vanished, and traditional reference points are now overcome. The composer doesn't feel the need to take into account the audience's expectations. What is created is bound to new writing procedures that don't necessarily have to be understood. Music thus enters an age of individuality but, at the same time, doesn't resolve itself in the stability of a new musical language. The 12-tone technique is a new paradigm, but not unalterable. It has nevertheless prompted an urge toward new approaches to composing born from and defined by reaction and contradiction (electroacoustic music, electronic music, concrete music, postmodern music). Music has followed the ways and the spirit of the time, the "Zeitgeist." Writing music has always strived toward constantly more open directions, depending on the methods of the immediate predecessors and the new technological means available.

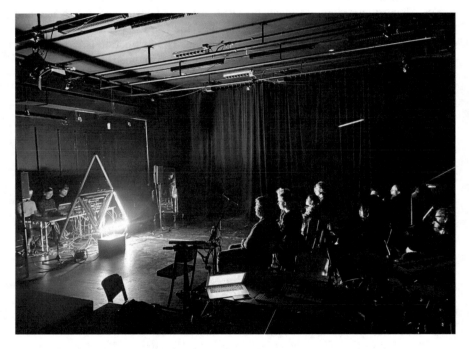

Photo: "Sound and visions," courtesy of Alexander Schubert

Box 1.2
According to an evolutionist concept, over time, musical language has followed a path that has gone from ditonic scales to the heptatonic scale, where it has found its final stability. From here evolution would have coincided with the development of natural resonance: from the seventh of Renaissance music to the eleventh of Debussy and then to the twelfth of Messiaen. But, with the

appearance of new expressive means, e.g., the use of electroacoustic instruments, natural resonance laws are undermined, and the concept of music expands. It is not only rhythm or pitch anymore, it is a sonic entity and sometimes noise. Aesthetics and listening attitudes necessarily have to change, hand in hand with the cultural and historical climate they have emerged from.

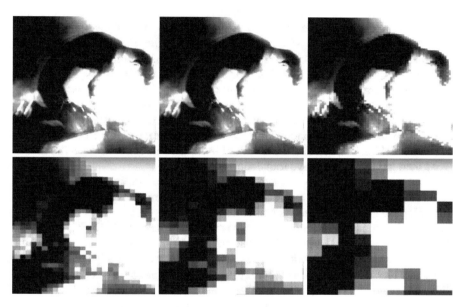

Photo: "Deconstructing sounds," courtesy of Lorenzo Colombo

1.2   The Shape of Music

The word "music" originally referred to the Muses and their art. It didn't have a specific meaning, but denoted all sorts of activities inspired by pleasure, well-being, and beauty. In mythology, the Muses were the nine daughters of Mnemosyne (that is Memory) and Zeus. They each had specific skills around thought (eloquence, persuasion) and singing in all its forms. Greek mythology tells about a very strong connection and an elective affinity between mind and music, thought and singing, and memories and melody. Moreover, the word "muse" could derive from an Indo-Germanic root "ma(n)," which means knowledge and thought: similar to the Italian word "mente", the Latin "mens" and the Sanskrit "manas" and the English word "mind". Clearly, music has always been recognized as a noble art, capable of putting in communication and in relation mind, spirit, and emotions.

In ancient times, music was seen as a way of redeeming: Pythagoras and his disciples used medicine to purify the body and music to achieve the same goal, but for the soul. According to the Pythagoreans, the universe followed the same rules that regulated the world of sound, and this was a sign that the essence of music was pure and ascetical. Furthermore, still in the Pythagorean school, it was recognized that pleasant sounds (consonant) were produced when the frequencies of two vibrating entities formed simple integer ratios (i.e., 2:1, the octave). On the contrary, complex ratios produced rough sounding tones (i.e., 16:15, dissonant minor second). The concept of musical purity resonates in the last century as well. One needs only think of what Jean Cocteau wrote in 1918 in *Le Coq et l'Arlequin*, taking as a point of reference for pure music the whole of Satie's work, followed by Webern's figurations and the 12-tone serialism [11]. These require a search for abstraction that is formal and rigorous and so also pure. Nevertheless, one can often perceive in the listener of such scores, a feeling of bewilderment and incomprehension. The atonal and often dissonant codex seems full of impurities and is not only difficult to approach but also a source of anguish and awkwardness, unpleasantness, and instability. For this reason, consonant pitch relationships occur more frequently in tonal music than in dissonant relationships. Moreover, a preference for consonance is observed early in life, well before an infant is exposed to culturally specific music. Does then tonal harmony have foundations?

Does our brain search for musical purity or is it already designed to discriminate between harmony (purity) and dissonance/atonality (impurity)? Classical neurophysiological studies have enabled us to compare the perception of consonant intervals (relationships of frequency that can be simple as octave, fifth, fourth, and major third, or complex as major seventh or dissonant as minor second or tritone) using the scalp-recorded frequency-following response (FRR), assuming that phase-locked neural activity within the brainstem may preserve information about consonant and dissonant pitch relationships. The results of these experiments confirm this fascinating idea: the strength of aggregate neural activity within the brainstem appears to be correlated with the relative ordering of consonance found behaviorally. Furthermore, basic properties of musical pitch structure are encoded both in subjects classified as non-musician and without musical ear training, and possessors of an absolute/perfect pitch. Intervals elicit graded levels of pitch salience and lead to different subjective ratings, accounting for the hierarchical ranking of musical intervals. This study confirms that the subcortical response to consonant pitch intervals generates more robust and synchronous phase locking than dissonant pitch intervals, implying that the perceptual allocation related to the preference of consonance could be a by-product of innate sensory-level processing [12]. Similar conclusions were obtained in behavioral studies in children: newborns prefer listening to consonant rather than dissonant musical sequences and don't like atonal melodies if compared to tonal songs. In particular, functional MRI was used to measure brain activity in 1- to 3-day-old babies while they heard excepts (21 s long, average tempo of 124 beats per minute) of Western tonal music (original piano music drawn from the corpus of major-minor tonal classical music by Bach, Mozart, and Schubert) and altered versions (with changes of tonal key or permanently dissonant) of the same tracks.

Music evoked predominantly right-hemispheric activations in primary and higher order auditory cortex (superior temporal gyrus, transverse temporal gyrus and planum polare, planum temporale and temporoparietal junction), confirming that music perception with hemispheric functional asymmetry (activation with right-hemispheric dominance) is present at birth. When presented with altered music, activations emerged in the left inferior frontal cortex and limbic structures. This pattern of activation presumably reflects a sensitivity to dissonance. These data corroborate the hypothesis that hierarchical representation of musical pitch is demonstrated at a subcortical level, suggesting that a listener's judgment of unpleasant (not pure) or pleasant (pure) sounding melody may be rooted in low-level sensory processing, governed by the fundamental capabilities of the auditory system. The perception of sensory dissonance is a function of the physical properties of auditory stimuli, as well as those of basic anatomical and physiological constraints, resulting from limitations of the auditory system in resolving tones that are proximal to pitch [13].

Moreover, it is interesting to observe how the analysis of the harmonic frequency of the human vocal spectrum (statistical structure of human speech) highlights a probability distribution with peaks that correspond to intervals of octaves, fourths, thirds, and major sixths, which are the intervals of the pentatonic scale. This suggests that the human ear could have an innate aesthetic and subjective preference for specific intervals, and this is hence linked to speech "sounds" (vocalizations) and to how the brain encodes them in the superior temporal cortex. So, humans prefer tone combinations that reflect the spectral relationships of human vocalization: across cultures, harmonic structure of speech closely parallels that of music, bearing in mind that human vowel sounds are based on the chromatic scale.

Are we consequently predisposed to take pleasure in harmony? Still, we are also fascinated by birdsong and the chirping of cicadas. After all, are these only sounds or do we find in them a musical coloring that lets us appreciate them on an emotional level? Why do animals make sounds? Many animals have sound-producing and sound-processing abilities. They process sound as it occurs within a time, ascribe a sort of meaning to this sound, and adjust their behavior accordingly.

Let's thing of the cricket. It produces distinct sounds thanks to the movement of its elytra, as a result of specific muscle activities and a system we can regard as musical. The phrase is peculiar to each species and is typically composed of five notes alternating with trills and sounds we could define as "cri-cri." This music serves as a sexual call for the species. A female does not react to the chirping of a male belonging to another species.

Thanks to a specialized phonate system named syrinx, birds can produce cries (of danger or as a call), whistles (sounds with a stable frequency within a time), and also actual songs. Among the nearly 9000 species of existing birds, at least 40% can elaborate harmonies: organized sounds made of specific note pitches, frequency variations even within the same note, and intervals between phrases. The cuckoo, for example, executes a song that is always characterized by the same interval, a third. When examining more complex elaborations, the case of the white wagtail (*Motacilla alba*) is paradigmatic as it produces an articulated melody, entirely individual to the species, composed first of a whistle, then of some modulated notes clustered in two types of syllables, and finally of a kind of humming, different from the initial whistle. In this species, we even find specific "dialects," different sonorous articulations depending on the region the birds are found. This means that there is the possibility that also animals can learn musicality and evolve their singing. Birdsong has a social and utilitarian function. It serves to distinguish between family members and strangers, to demarcate a territory, and to promote sexual activity. Birdsong also follows yearly, or even daily, rhythms and is a kind of signal or program.

So, does music have a sort of adaptive function also in humans? Does it enhance cooperation, social cohesion, and assistance within a group? Is there a sort of continuity across species whereby musical sounds are able to influence physiological processes that increase physical and mental well-being?

It is in fact likely that a possible reason that music arises and endures is because it brings people together. Music definitely has a function in social cohesion. Music can be used for communication and has a ritual significance in every religion.

From an evolutionary point of view, spoken language and music evolved from a proto-language driven by gesture, framed by musicality, and performed thanks to the flexibility that accrued with continuous anatomical developments, not only in the brain, but also in the coordination of our laryngeal, pharyngeal, and facial muscles. So, as cultures evolves from simple to complex, moving from primitive to civilized, music evolves from simple to complex within societies as they progress. In this context, recent studies confirm that music is able to activate both phylogenetically ancient reward/valuation brain systems (striatal, limbic, and paralimbic) and more evolute perception/prediction systems (temporofrontal) [14]. This pattern may be interpreted as a possible indicator that the human brain holds an adaptive neural specialization for processing music as a rewarding stimulus, although only with an aesthetic value. Music for pleasure, then, is one of the hallmarks of what it means to be a human being.

Photo: "The art ensemble," courtesy of Guido Borso

Acknowledgments The author gratefully thanks Dr. Francesca Ferretti for the friendly and enthusiastic help.

References

1. www.Britannica.com/art/music
2. Rousseau JJ. In: Diderot D, D'Alembert J, editors. Encyclopedie ou dictionnaire raisonne' des sciences, des arts et de mestieres. Paris: Briasson; 1750.
3. Adorno T. Philosophie der neuen music. Frankfurt am Main: Suhrkamp; 1958.
4. Xenakis I. Musique, architecture. Paris: Castermann; 1971.
5. Cage J. Pour les oiseaux, entretiens avec Daniel Charles. Paris: Belfond; 1976.
6. Charles D. Esercizi di silenzio. In: De Melo Pimenta ED, editor. John Cage, il silenzio della musica. Silvana Editore: Cinisello Balsamo; 2003.
7. Ellis AJ. On the musical scales of various nations. J Soc Arts. 1885;XXXIII:458–527.
8. Merriani AP. The anthropology of music. Enanston Ill: Northwestern University Press; 1964.
9. Levi SC. Le cru et le cult. Paris: Plon; 1964.
10. Schoenberg A, Busoni F. Briefe. Beitrage zur musikwissenschaft. 1977;XIX, 3:173–78.
11. Le CJ. coq et l'Arlequin. Paris: Editions de la Sirene; 1918.
12. Bidelman GM, Krishnan A. Neural correlates of consonance, dissonance and the hierarchy of musical pitch in the human brainstem. J Neurosci. 2009;29(42):13165–71.
13. Perani D, Saccuman MC, Scifo P, et al. Functional specializations for music processing in the human brain. PNS. 2010;107(10):4758–63.
14. Edwards RD. The neurosciences and music education: an online database of brain imaging neuromusical research. 2008. http://library.uncg.edu.

Synesthesia and Emotional Sound

2

Lorenzo Lorusso, Amy Ione, Antonia Francesca Franchini, and Alessandro Porro

2.1 Introduction

The etymology of synesthesia is derived from the Greek *syn* meaning "together" and *aisthesis* meaning "sensation" or "perception." Often translated as "a union of the senses," the concept expresses the idea of combined or simultaneous perception [1, 2]. Aisthesis was the term used by the philosopher Aristotle (384 BC–322 BC) in the fourth century BC and later by physicians such as Soranus of Ephesus and Aretaeus of Cappadocia, both of the second century AD. Three ways of interpreting synesthesia are discussed in this paper: rhetorical, developmental, and medical.

The first, rhetorical, metaphoric, or stylistic, unifies different sensory spheres linguistically or imagistically. Before scientific investigations ensued, it was frequently assumed that all references to synesthesia fell into this category. A painting might depict visual music, or language combinations might evoke sensory experiences through references to bright colors, warm voices, or clear sounds.

Constitutional or developmental synesthesia, the second type discussed below, refers to a genetic trait found in a subset of normal people. The literature has identified at least 60–200 subtypes of synesthesia that fall into this category [3]. The condition occurs in about 4% of the general population [4, 5]. There is evidence that it involves genetics because it has been shown to run in families, most commonly from mother to daughter (synesthesia is more prevalent among females) [6]. The most common form of synesthetes in this category is grapheme-color synesthesia

L. Lorusso (✉)
U.O.C. Neurologia & Stroke, A.S.S.T.-Lecco, Merate (LC), Italy

A. Ione
Diatrope Institute, Berkeley, CA, USA

A. F. Franchini · A. Porro
Dipartimento di Scienze Cliniche e di Comunità e CRC Centro di Salute Ambientale, Università degli Studi di Milano, Milan, Italy

© The Author(s), under exclusive license to Springer Nature Switzerland AG 2022
B. Colombo (ed.), *The Musical Neurons*, Neurocultural Health and Wellbeing, https://doi.org/10.1007/978-3-031-08132-3_2

(64%). The second form is time unit (e.g., names of days or months) color synesthesia (22%), followed by musical sound color synesthesia (18%) [7]. Scientific investigators of this benign, alternative form of perception began to examine the possibility of genetic synesthesia related to cognition and physiology in the nineteenth century. Present from an early age, constitutional or developmental synesthesia does not change significantly during one's lifetime. In this form, stimulation of one sensory or cognitive pathway leads to automatic, involuntary experiences in a second sensory or cognitive pathway.

The third type, medical synesthesia, refers to a form that is either a neurological disorder or comes about following psychotropic drug ingestion. It occurs independently and inadvertently during adulthood as a result of pathological conditions. In contrast to its developmental counterpart, this type of synesthesia does not demonstrate either idiosyncrasy or automaticity [8–10].

Emotional reactions play a part in the synesthetic process. Additionally, cerebral structure processing emotions are differents in developmental synesthesia and in the acquired form [11, 12]. Given this and the research that demonstrates that all three types of synesthesia link to music perception [13], we hypothesize that forms of synesthesia that include music potentially have an influence on well-being.

2.2   Historical Foundations and Philosophical Debates

Natural philosophy, a precursor to modern science, included several foundational ideas about perception and sensory experience. Aristotle (384 BC–322 BC), for example, wrote that the harmony of colors was like the harmony of sound describing the five senses distinctly [14–16]. Although he additionally emphasized that the various senses can interact with each other creating different sensory modalities, his view is not the kind of genetic synesthesia now discussed in the constitutional and developmental literature. This is because at that time a joining of the senses, or synesthesia, was largely considered neurologically abnormal because it was at odds with the five distinct senses Aristotle notably codified [17].

Further evaluation of the relationships among the senses was an issue in the seventeenth century in the work of John Locke (1632–1704), particularly "An Essay Concerning Human Understanding" published in 1690 [18]. Locke grappled with a question raised by the Irish natural philosopher William Molyneux (1656–1698) [19] on the role of the senses regarding human knowledge: he had asked if a blind man, who recognizes objects by touch, would be able to recognize and define them by sight alone when he suddenly regains that sense. In particular, he raised the question of the hierarchy of the senses and the translatability of sensory data among them [20]. Others, such as Johann Gottfried Herder (1744–1803), a pupil of Immanuel Kant (1724–1804), came to a more active and complex conception of the human sensory structure, seeing it as a central feature of the human senses. Its "meaningfulness" (*Besonnenheit*) is evident because it allows one to become aware of the immediate recognition of objects.

Just as Locke and Herder disagreed, there were contrasting color theories, largely discussed in relation to the senses and harmony. The idea of harmony within the Baroque era (seventeenth and eighteenth centuries) was built on ideas that equated color harmony with the spheres by Johannes Kepler (1571–1630) and the color theories of Athanasius Kircher (1601–1680) and Isaac Newton (1642–1727).

Newton, in the 1660s, had the opportunity in Cambridge to organize the results of his earlier optical research and follow-up on his interests in Descartes' work, including *Géométrie* [21]. Shortly after his 1672 election to the Royal Society, Newton presented his first public paper, a controversial study on the nature of color. In this paper, he also investigated the coincidence of the seven colors of the spectrum with intervals of a musical scale [22]. René Descartes (1596–1650) gave a scientific explanation for the rainbow's formation in 1637 [23] and Newton added color to Descartes' arc. In 1704, he created the disc that bears his name (based on the idea of natural equivalence between the spectrum of colors and musical notes).

His main detractor, Johann Wolfgang von Goethe (1749–1832) [24], pointed out that color depends not only on light but also on our senses and how we perceive them [25]. He realized that colors are the product of the visual system and do not depend only on the light that reaches the eye.

Goethe's views on color influenced nineteenth-century painters, especially the Impressionists. Many modern paintings refer to the pure abstraction of musical training processes and impose an essential consideration of relations and interferences between music and painting. The distinction between seeing and hearing, as relevant from the physiological point of view, has never been enough to prevent mutual interaction. One of the most well-known interactions between eye and ear, the so-called "hearing colors" (colored hearing), has been a vast subject of numerous experiments and observations [26, 27].

Synesthesia as a specific sensory feature of human nature also played a part in philosophical anthropology for the first half of the twentieth century, thanks to Max Scheler (1874–1928) and Arnold Gehlen (1904–1976). Their fundamental assumptions bind human nature and multisensory perception: the first being phylogenetic, which distinguishes men from other animals through their morphology, and the second ontogenetic allowing men to distinguish themselves through multi-sensorial plasticity.

2.3   Scientific Development of the Term Synesthesia

Rare and anecdotal cases of synesthesia described before the nineteenth century show that the phenomenon was not well understood [28–30]. Modern scientific studies began to emerge in the late eighteenth and early nineteenth centuries as the neurosciences studied the relationship between the mind and the human nervous system [31–33]. In 1812, there was an initial description of the phenomenon of synesthesia by Georg Ludwig Sachs (1786–1814), who was suffering from albinism, as was his sister [34–36]. In Italy [37], the first descriptions of the relationship between sounds and colors were those of Carlo Botta (1766–1837), who in 1801 published an essay on the individual ability to associate sounds and colors based on the theory of Newton's color disc [38, 39]. In 1848, the Swiss ophthalmologist Charles Auguste Édouard Cornaz (1825–1911) considered the Sachs case in his doctoral thesis entitled "Des yeux des abnormalités congénitales et de leur annexes," concluding it was a common, inherited disorder afflicting males, characterized by an alteration of color perception, which he called hyperchromatopsia or "alteration du sens de la vue." In 1863, M. Perroud proposed that the perceptual disturbance was due to a psychological defect, following the ideas of the experimental psychology of the time. A year later, D. Chabalier changed the term hyperchromatopsia to pseudochromestesia, confirming psychological theories such as a "trouble des idées", or mental disorders such as the false association of ideas [40]. In 1864, Andrea Verga (1811–1895), quoting Chabalier, considered pseudochromestesia as an alteration of visual perception or a possible mental disorder based on a false association of ideas [41]. The debate was enriched by the contribution of Antonio Berti (1812–1879), who considered the disorder as having a psychic origin, characterized by a disturbance of memory and not connected to a disturbance of visual perception [42] and called it dyschromesthesia [43]. The pre-synesthetic debate ended with Berti, whose work on the issues of visual perception and the psychological component raised interest among French and Italian neuro-psychiatric schools.

The word synesthesia also made its appearance in 1865 in both France and Italy. In France, Alfred Vulpian (1826–1887) used the term. The Italian physiologist Filippo Lussana (1820–1897) [44–47] coined the word in the same year, using the term to describe the relationship between sounds and colors and connecting them to an alteration of the nervous centers of the brain [48]. Lussana, influenced by the theories of brain localization, from his point of view – still weakly related to phrenology – suggested that the centers for "color" and "melody" were located in the frontal convolution. He contributed to asserting that synesthesia is a neurological phenomenon: the brain's ability to take action for the integration of the various sensory perceptions. Lussana's research on musical notes led him, among other things, to also take an interest in the anatomy and physiology of the inner ear, contributing to a better understanding of how it works [48–50]. In 1873, J. A. [recte F. A.] Nussbaumer (1848–post 1887) added to the debate with a description of two brothers affected by synesthesia using a questionnaire, the first reported in history [40]. Later researchers, such as Gustav Theodor Fechner (1801–1887) in 1876, the psychiatrist Eugene Bleuler (1857–1939), along with his student K. Lehmann in 1881, and then Francis Galton in 1883, carried out epidemiological studies to quantify the phenomenon in the population and found that it had a certain distribution [51, 52]. Bleuler found that 12% of 600 patients had synesthesia. Using the recently developed psychoanalytic theories of Sigmund Freud (1856–1939), he concluded that synesthesia was a phenomenon characterized by a disorganized, widespread abnormal perception in schizophrenic patients [52, 53].

Cesare Lombroso (1835–1909) and Aleksandr Romanovich Luria (1902–1977) studied synesthetic phenomenon from an anthropological and psychological standpoint in the late nineteenth century. Lombroso's book "The Man of Genius" (Appendix VI) [54] mentioned colored hearing and reported a presentation by

Edouard Gruber [55]. Lombroso classified colored hearing in his theories on heredity, degeneration, and atavism. He considered it to be a pathology characterized by excesses, such as polydactyly. Synesthetes mentioned include the physicist and philosopher Fechner, a key figure in the founding of psychophysics, and the poet Arthur Rimbaud (1854–1891) [56]. The word was more widespread at the Congress of Physiological Psychology (1889), where it was associated with the term *audition colorée* "colored hearing" [57].

In 1892, Jules Millet introduced the term synesthesia in his thesis, together with that of "colored hearing" (audition colorée). In addition, in 1898 Jean Clavière [58] printed a remarkable bibliography. In the twentieth century, Luria studied Solomon Venianinovič Šereševskij, a young synesthete, for decades. Luria's observations were published in a small book, *The Mind of a Mnemonist: A Little Book About a Vast Memory* [59, 60]. This work's methodological importance stems from its concern with the relationship between super memory and synesthesia, also emphasized by Macdonald Critchley (1900–1997) [61]. These authors argued for a superior memory in synesthetes, with a study confirming this possible aspect [62]. However, it remains unclear as to what extent superior memory is a characteristic of all synesthetes and Luria's case focused on that superior memory rather than the fact that the subject had synesthesia [63]. In 1976, Luria highlighted the importance of the Šereševskij case as a paradigm for a scientific non-reductionist view, having been defined by the Russian psychologist in terms of romantic science [64].

2.4   Science, Romanticism, and Synesthesia

A noteworthy quality of the early scientific research was the difficulty in engaging quantitatively with the synesthetic experience. The work of Théodore Flournoy (1854–1920), a Professor of Psychophysiology, shows this well. A student of William Wundt and a friend of William James, Flournoy was a part of the rising movement striving to establish psychology's legitimacy as a science [65]. His synesthesia research, which endeavored to bring basic sensations, consciousness, and ideas together, resisted the kind of experimental testing practiced by Wundt because of the difficulty in devising quantifiable, independent measures for examining the experience. This led him to employ discursive introspective methods that were closer to William James' methods [65]. Although he relied on subjective data, his case studies on synesthetic personification are now considered pioneering and insightful.

By contrast, the romantic poets and writers considered synesthesia their manifesto. They saw the phenomenon as a tool in capturing the elements of nature, allowing them to recognize the various correspondences between all the senses and to unravel the mysteries that are hidden behind appearances. For example, Charles Baudelaire (1821–1867), in his essay "Writings on Art" [66], examined the relationship between color, sound, and smell. Baudelaire was not the main architect of this *raison d'être*; a group of poets and writers also adopted this view. These included Julie Théophile Gautier (1811–1872), Arthur Rimbaud, and Pierre Joris-Karl

Fig. 2.1 Some musician
synesthetes (public
domain)

Alexander Nikolaevič   Amy Beach (1867-1944)
Skrjabin (1872-1915)

Duke Ellington   John Cage (1912-1992)
(1899-1974)

Husmans (1848–1907) [67]. The list of artists associated with synesthesia in the nineteenth and twentieth century is impressive. Musicians include Franz Liszt, Alexander Laszlo, Jean Sibelius, Alexander Scriabin, Amy Beach, Duke Ellington, John Cage, and Olivier Messiaen [68–73] (Fig. 2.1).

Examples in the visual arts include Vassily Kandinski, Johannes Itten, and the abstract painter Luigi Veronesi. Painter-musicians such as Mikalojous Konstantinas Čiurlionis are also included [74, 75]. For them it was essential to listen to their inner-self. Their reasoning was that the five senses were rooted in the soul and served as a foundation for artistic creativity. For these artists, a combination of certain colors is compared to combinations of different tones. When in a proper agreement, an intensification of a tonal coloring results. This concept is mirrored in the field of music, as with modern orchestration, to which Richard Wagner (1813–1883) and Claude Debussy (1862–1918) contributed. This musical sensitivity developed and spread, involving different forms and genres of music that led to synesthetic representations of various kinds, such as jazz and pop [76, 77]. Like Flournoy's early case study work, the artist reports are based on qualitative data. Therefore, we can neither definitively validate any reports nor separate metaphoric examples from genetic ones. This validation problem is one reason scientific interest waned until objective measures of testing were developed late in the twentieth century.

2.5   Neurodevelopmental and Acquired Pathological Synesthesia

Synesthesia has only recently been seen as a pathological condition in neurodevelopmental disorders like Asperger's syndrome and Williams syndrome [78, 79]. While we characterize pathological synesthesia, which is acquired or adulthood variety [12] as a separate type here, it is prudent to note that this is largely because recent research has refined our understanding of the synesthete experience. With the range of current research in mind, it is also important to note that historical papers did identify pathologically acquired synesthesia in epilepsy, migraine, and in other neurological disorders, so there is a connection between recent work.

Oliver Sacks (1933–2015), in his book *Hallucinations,* mentions that the first pathological synesthetic phenomena were described in 1881 by William Richard Gowers (1845–1915), who in his essay *Epilepsy* gave various examples of patients affected with epilepsy who spoke of hearing "the sound of a drum," "hissing," "ringing," and "rustling" as a hyperphysiological state. In the same period, other clinicians reported cases affected by epilepsy, such as the Scottish David Ferrier (1843–1928), who worked in London and mentioned an epileptic patient with a remarkable synesthetic aura, in which she would experience "a smell like that of green thunder" [80–82]. John Hughlings Jackson (1835–1911) reported on several cases, the first in 1863, describing a woman with colored vision associated with epilepsy [83]. Other clinical cases in the following years were reported by Jackson and classified with the name of "uncinate group of fits" due to different causes, as confirmed by various researchers [84–86]. Neckar and Bob demonstrated that a specific synesthetic-like mechanism could increase temporal-limbic excitability [87].

On the other hand, little has been published on synesthesia occurring as a migraine aura symptom [88, 89]. In 1939, Critchley proposed that aura and confusional states might occur in migraine and epilepsy as synesthetic phenomena [81]. Sacks and Podoll described patients suffering from migraine with aura and considered their attack as a synesthetic equivalence between auditory stimuli and visual image [90–92]. Based on the prevalence in the general population, about 2 in 1000 migraineurs would be affected by some kind of synesthesia, most of them of the visual type [89, 93]. The patho-mechanisms of auditory-visual synesthesia in migraine are obscure.

Synesthetic episodes are reported in patients after a thalamic ischemic stroke as well. Studies demonstrate a double dissociation in the patient's secondary somatosensory cortex (increased response to auditory stimulation and decreased responses to somatosensory stimulation) and suggest that stroke insult-induced plasticity can result in abnormal connections between sensory modalities that are normally separate, and synesthesia can be caused by inappropriate connections between nearby cortical territories [10, 94, 95]. In the last 50 years, other neurological disorders have been described in association with synesthesia, such as in retinitis pigmentosa [96, 97] and in the presence of gliotic mass [96, 98] with unclear explanation on their mechanism [96]. Other cases have been reported in association with irritable bowel syndrome and a possible involvement in multiple sclerosis with a plausible role of the immune system in the nondevelopmental synesthesia [99–101].

Since the first studies by Bleuler, few have addressed if synesthesia is linked to more widespread abnormalities in perception in neuropsychiatric subjects in understanding the mechanisms that mediate the development of a possible typical and atypical perceptual experience [52, 53, 102]. On the other hand, since the beginning of scientific research on psychedelic drugs [103], numerous observations attest that a wide range of chemical substances elicit synesthesia with a possible implication of the serotonergic system [104, 105]. We need further studies on developmental and acquired pathological conditions to understand the neurophysiological and neurochemical mechanisms underlying the synesthesia phenomenon.

2.6   Music Synesthesia and a Possible Influence on Well-being

Different studies investigated the capacities of people to associate sound synesthesia with colors, with a strong correlation between the emotional associations of the music and those of the colors chosen to go with the music [106–110]. Through this work, the association of sound synesthesia and emotion with pleasant perceptions has recently emerged, although some experts have raised questions about this relationship, noting unpleasant perceptions of some musician synesthetes [111, 112].

The pleasurable response is present in all stages of childhood and adulthood [113]. Researchers have also found that in later life the synesthetic colors are less chromatics [114, 115], and with particular personality traits and cognitive characteristics, there is a significant relationship with anxiety [116] and mood [117–119]. Factors to consider include the type of music and which instruments in an orchestra stimulates the synesthetic emotion/sensation to determine environmental properties [120, 121].

Fruitful investigations have probed how pleasure, associated with sound in the form of music or noise, affects taste. Various studies have demonstrated the presence of sound-gustatory synesthesia by looking at the possible effects of background music on the perceived taste in different settings. In a coffeehouse, sweetness was perceived more while listening to the "slow" music. The perception of sourness increased with the tempo of the music track, even when sour and salty components were not present in their drinks. There was a sensation of bitterness when normal and fast music was played; sourness and saltiness, and the perception of sourness increased with the speed of the music track [122–125]. Also, emotional synesthetic percepts were described in acquired cerebral disorder in a patient who had a posterolateral thalamus hemorrhage. This patient experienced blue photism, intense extracorporeal sensation when hearing brass instruments [10, 126]. In addition, color association with music is prevalent in our society and every day sounds can trigger emotions and feeling of well-being with involvement of the left anterior insular cortex to construct a dynamic representation of the current state of emotional well-being [127–129].

Music listening may offer a means to better mediate musical understanding and meaning. As such, it serves as a potentially valuable tool for music educators to explore their students' musical understanding and imagination and nurture their well-being [113, 130, 131]. Digital tools can assist in this. They can play an important role in creating the types of emotions associated with sound synesthesia, allowing more accessibility to the experience. For example, through interaction with video in synesthetic ways, we can stimulate and support user creativity. We can also use digital tools to learn more about the synesthesia experience through monitoring how videos influence and are influenced by users and we can study the ambient when they are being played [132, 133].

2.7 Conclusion

Over time our understanding of human physiology and the physiology of perception has grown and changed conclusions about sensory modalities. We now know that synesthesia is probably more diffuse than formerly thought. Recent research has also shown that synesthetic experiences take many forms. In addition, the wide variety of sensory reports has added fascinating avenues for multidisciplinary studies with a possible benefit on our well-being.

References

1. Marcovecchio E. Dizionario etimologico storico dei termini medici. Festina Lente: Impruneta; 1993. p. 839.
2. Cytowic RE. Synesthesia. In: A Union of the Senses. Cambridge: MIT Press; 2002.
3. Day SA. Types of synaesthesia; 2021. http://www.daysyn.com/types-of-syn.html. Accessed 21 Dec 2021.
4. Cytowic RE, Eagleman DM. Wednesday is indigo blue: discovering the brain of synesthesia. Cambridge: MIT Press; 2009.
5. Day SA. Synesthesia. In: Demographic aspects of synesthesia; 2021. http://www.daysyn.com/types-of-syn.html. Accessed 21 Dec 2021.
6. Hubbard EM, Ramachandran VS. Neurocognitive mechanisms of synesthesia. Neuron. 2005;48:509–20.
7. Day S. Some demographic and socio-cultural aspects of synesthesia. In: Robertson LC, Sagiv N, editors. Synesthesia: perspective from cognitive neuroscience. New York: Oxford University Press; 2005. p. 11–3.
8. Harrison J. Synaesthesia. The strangest thing. Oxford: Oxford University Press; 2007.
9. Rogowska A. Categorization of synaesthesia. Rev Gen Psychol. 2011;15:213–27.
10. Schweizer TA, Li Z, Fischer CE, Alexander MP, Smith SD, Graham SJ, Fornazzarri L. From the thalamus with love, a rare window into the locus of emotional synesthesia. Neurology. 2013;81:509–10.
11. Perry A, Henik A. The emotional valence of a conflict: implications from synesthesia. Front Psychol. 2013;4:978.
12. Safran AB, Sanda N. Color synaesthesia. Insight into perception, emotion, and consciousness. Curr Opin Neurol. 2015;28:36–44.
13. Bragança GFF, Fonseca JGM, Caramelli P. Synesthesia and music perception. Dement Neuropsychol. 2015;9:16–23.

14. Tornitore T. Storia delle sinestesie. Le origini dell'audizione colorata. Genova: Brigati-Carucci; 1986.
15. Manzoni T. Aristotele e il cervello. Le teorie del più grande biologo dell'antichità nella storia del pensiero scientifico. Roma: Carocci; 2007.
16. Day SA. Synesthesia resources; 2021. http://www.daysyn.com/types-of-syn.html. Accessed 07 Jan 2022.
17. Ione A, Tyler CW. Neuroscience, history and the arts. Synesthesia: is F-sharp colored violet? J Hist Neurosci. 2004;13:58–65.
18. Locke J. An essay concerning human understanding. London: Printed [by Elizabeth Holt] for Tho. Basset, and sold by Edw. Mory at the sign of the three bibles in St. Paul's Churchyard; 1690.
19. Mazzeo M. Storia naturale della sinestesia. Dalla questione Molyneux a Jakobson. Macerata: Quodlibet; 2005.
20. Morgan M. Molyneux's question: vision, touch and the philosophy of perception. Cambridge: Cambridge University Press; 1977.
21. Descartes R. Géométrie. Leyde: De L'imprimeterie de Ian Marie; 1637.
22. Newton I. New theory about light and colours. Phil Trans R S. 1671–1672;80:3075–87.
23. Descartes R. Discours de la méthode. Leyde: De L'imprimeterie de Ian Marie; 1637.
24. Judd DB. Introduction. In: Goethe JG, editor. Goethe's theory of colours. Translated from the German with notes by Charles Locke Eastlake. Cambridge, Massachusetts and London: The MIT Press; 1970. p. vii–viii.
25. Goethe W. Beiträge zur Optik. Zur Farbenlehre. Weimar: Industrie-Comptoir; 1791–1792.
26. Haigh A, Brown DJ, Meijer P, Proulx MJ. How well do you see what you hear? The acuity of visual-to-auditory sensory substitution. Front Psychol. 2013; https://doi.org/10.3389/fpysg.2013.00330.
27. Harvey JP. Sensory perception: lessons from synesthesia. Using synesthesia to inform the understanding of sensory perception. Yale J Biol Med. 2013;86:203–16.
28. Larner AJ. A possible account of synaesthesia dating from the seventeenth century. J Hist Neurosci. 2006;15:245–9.
29. Jewanski J, Simner J, Day SA, Ward J. The development of a scientific understanding of synesthesia from early case studies (1849-1873). J Hist Neurosci. 2011;20:284–305.
30. Jewanski J. Synesthesia in the nineteenth century: scientific origins. In: Simner J, Hubbard EM, editors. Oxford handbook of synesthesia. Oxford: Oxford University Press; 2014. p. 369–98.
31. Clarke E, Jacyna LS. Nineteenth-century origins of neuroscientific concepts. Berkeley: University of California Press; 1987.
32. Ione A. Art and the brain: embodiment, plasticity, and the unclosed circle. Amsterdam: Rodopi Brill; 2016.
33. Jewanski J, Simner J, Day SA, Rothen N, Ward J. The evolution of the concept of synesthesia in the nineteenth century as revealed through the history of its name. J Hist Neurosc. 2019; https://doi.org/10.1080/0964704X.2019.1675422.
34. Sachs GTL. Historia naturalis duorum leucaetiopum auctoris ipsius et sororis eius. Solisbaci: Sumptibus Bibliopolii Seideliani; 1812.
35. Lima ESC. Cross-sensory experiences and the enlightenment: music synesthesia in context. Música Hodie. 2019;19:e54919.
36. Jewanski J, Simner J, Day SA, Rothen N, Ward J. The "golden age" of synesthesia inquiry in the late nineteenth century (1876–1895). J Hist Neurosci. 2020;29(2):175–202. https://doi.org/10.1080/0964704X.2019.1636348.
37. Lorusso L, Porro A. Coloured-synaesthesia in 19th century Italy. In: Clifford Rose F, editor. Neurology of music. London: Imperial College; 2010. p. 239–56.
38. Botta C. Memoire sur la nature des tons et des sons. Mémoires de l'Académie des Sciences, Littérature et Beaux-Arts de Turin pour les années X et XI. 1801;12:191–214.
39. Botta C. Scritti minori. Biella: Amosso; 1860. p. 17–39.

40. Jewanski J, Day SA, Simner J, Ward J. The beginnings of an interdisciplinary study of synaesthesia: discussions about the Nussbaumer brothers (1873). Theoria et Historia Scientiarum. 2013;10:149–76.
41. Verga A. La pseudocromestesia. Gazzetta Medica Italiana-Lombardia. 1864;49:426–7.
42. Zago S, Randazzo C. Antonio Berti and early history of aphasia in Italy. Neurol Sci. 2006;27:449–52.
43. Berti A. Della pseudocromestesia. Archivio Italiano per le malattie Nervose. 1865;2:22–8.
44. Zanchin G, Lisotto C, Maggioni F. Filippo Lussana (1820-1897), a physiologist of the Paduan medical faculty and his contribution to neurology. Italian J Neurol Sci (Suppl) Cogito. 1992;23:79–84.
45. Bock G. Lussana Filippo. In: Dizionario Biografico degli Italiani, vol. 61. Roma: Istituto dell'Enciclopedia Italiana; 2006. p. 269–72.
46. Berbenni G, Lorusso L. Filippo Lussana (1820–1897) da Cenate alle neuroscienze. Atti dell'incontro di studio, Cenate di Sopra, 26 maggio 2007. Bergamo: Fondazione per la storia economica e sociale di Bergamo; 2008.
47. Lorusso L, Bravi GO, Buzzetti S, Porro A. Filippo Lussana (1820–1897): from medical practitioner to neuroscience. Neurol Sci. 2012;33:703–8.
48. Lussana F. Lettera seconda. Fisiologia morale dei colori. Archivio Italiano per le malattie nervose. 1865;2:141–8.
49. Lussana F. Fisiologia dei colori. Padova: Sacchetto; 1873.
50. Lussana F. Sull'udizione colorata. Archivio Italiano per le malattie Nervose. 1884;21:371–7.
51. Fechner GT. Vorschule der Aesthetik. Leipzig: Bretkopf & Härtel; 1876.
52. Bleuler E, Lehmann K. Zwangsmässige Lichtempfindungen durch Schall und verwandte Erscheinungen auf dem Gebiete der andern Sinnesempfindungen. Leipzig: Fues; 1881.
53. Bleuler E. Zur Theorie der Sekundärempfindungen. Z Psychol. 1913;65:1–39.
54. Lombroso C. L'uomo di genio. Torino: Fratelli Bocca; 1894.
55. Galton F, Gruber E. L'audition colorée et les phenomenes similaires. In: Congrès international de psychologie expérimentale de Londres 1892. Londres: Eberchard et Jacot; 1893.
56. Lombroso C. L'uomo di genio. Roma: Napoleone; 1971.
57. Jewanski J, Simner J, Day SA, Rother N, Ward J. Recognizing synesthesia on the international stage: the first scientific symposium on synesthesia (at the international conference of physiological psychology, Paris, 1889). J Hist Neurosci. 2020;29(4):357–84.
58. Clavière J. L'audition colorée. Annee Psychol. 1898;5:161–78.
59. Luria AR. Malen'kaja knizka o bol'soj pamjati. Moskwa: Izdatel'stvo Moskowskogo Universiteta; 1968.
60. Luria AR. The mind of a mnemonist, a little book about a vast memory. New York & London: Basic Books; 1968.
61. Critchley M. In memoriam. In: Luria AR, editor. Brain and language, vol. 5; 1978. p. v–vi.
62. Mills CB, Innis J, Westendorf T, Owsianiecki L, McDonald A. Influence of a synesthete's photisms on name recall. Cortex. 2006;42:155–63.
63. Ward J, Mattingley JB. Synaestesia: an overwiev of contemporary findings and controversies. Cortex. 2006;42:129–36.
64. Luria AR. Romantic science: unimagined portraits. In: Luria AR, editor. (1979) Viaggio nella mente di un uomo che non dimenticava nulla. In appendice due scritti inediti. Roma: Armando; 1976. p. 108–16.
65. Plassart A, White RB. Theodore Flournoy on synesthetic personification. J Hist Neurosci. 2017;26:1–14. https://doi.org/10.1080/0964704x.2015.1077542.
66. Baudelaire C. Ecrits sur l'Art. Paris: Flammarion; 1863.
67. O'Malley G. Literature synaesthesia. J Aesthet Art Critic. 1957;15:391–411.
68. Ione A. Neurology, synaesthesia, and painting. Intern Rev Neurobiol. 2006;74:69–78.
69. Pearce JMS. Synaesthesia. Eur Neurol. 2007;57:120–4.
70. Triarhou LC. Neuromusicology or musiconeurology? "Omni-art" in Alexander Scriabin as a fount of ideas. Front Psychol. 2016; https://doi.org/10.3389/fpsyg.2016.00364.

71. Day SA. Regarding types of Synesthesia and color music art. In: Galeyev BM, editor. Synesthesia: common wealth of senses and synthesis of arts. Kazan', Russia: Izdatel'stvo KGTU; 2008. p. 282–8.
72. Kuo-Ying L. Cross-cultural perspectives of synesthesia in contemporary musical compositions. Int J Hum Art Soc Stud. 2020;1(1) https://doi.org/10.5121/ijhas.2020.01102.
73. Jewanski J, Day SA, Siddiq S, Haverkamp M, Reuter C, editors. Music and synesthesia. Abstracts from a conference in Vienna, scheduled for July 3–5, 2020. Münster: Wissenschaf tliche Schriften der WWU Münster; 2020.
74. Tedeschi F, Bolpagni P. Visioni musicali. Rapporti tra musica e arti visive nel novecento. Milano: Vita e Pensiero; 2009.
75. Vergo P. The music of painting. Music, modernism and the visual arts from the romantics to John cage. London: Phaidon Press Limited; 2010.
76. Merriam AP. Synestehsia and intersense modalities. In: Merriam AP, editor. The anthropology of music. Chicago: Northwestern University Press; 1964.
77. Moseley I. Crossed wires. Synaesthetic responses to music. In: Clifford Rose F, editor. Neurology of music. London: Imperial College; 2010. p. 257–75.
78. Neufeld J, Roy M, Zapf A, Sinker C, Emrich HM, Prox-Vagedes V, Dillo W, Zedler M. Is synesthesia more common in patients with Asperger syndrome? Front Hum Neurosci. 2013;7:847. https://doi.org/10.3389/fnhum.2013.00847.
79. Thornoton-Wells T, Cannistraci CJ, Anderson A, Kim C, Eapen M, Gore JC, Blake R, Dykens EM. Auditory attraction: activation of visual cortex by music and sound in Williams syndrome. Am J Intellect Dev Disabil. 2010;115:172–89. https://doi.org/10.1352/1944-7588-115.172.
80. Gowers WR. Epilepsy and other chronic convulsive diseases: their causes, symptoms and treatment. London: Churchill; 1881.
81. Critchley M. Neurological aspect of visual and auditory hallucinations. BMJ. 1939;2:634–9.
82. Sacks O. Hallucinations. New York: First Vintage Books Edition; 2013.
83. Jackson HJ. Epileptiform seizures— aura from the thumb—attacks of coloured vision. Med Times Gaz. 1863;1:589.
84. Jackson HJ. On right or left-side spasm at the onset of epileptic paroxysms, and on crude sensation warnings and elaborate mental states. Brain. 1880;3:192–206.
85. Jackson HJ, Stewart P. Epileptic attack with a warning of a crude sensation of smell and with the intellectual aura (dreamy state) in a patient who had symptoms pointing to gross organic disease of the right temporo-sphenoidal lobe. Brain. 1899;22:534–43.
86. Daly D. Uncinate fits. Neurology. 1958;8:250–60.
87. Neckar M, Bob P. Synesthetic association and psychosensory symptoms of temporal epilepsy. Neuropsychiatr Dis Treat. 2016;12:109–12.
88. Podoll K, Robinson D. Auditory-visual synaesthesia in a patient with basilar migraine. J Neurol. 2002;249:476–7.
89. Alstadhaug KB, Benjsminsen E. Synesthesia and migrane. BMC Neurol. 2010;10:121. https://doi.org/10.1186/1471-2377-10-121.
90. Sacks OW. Migraine. Revised and expanded. Berkeley-Los Angeles-Oxford: University of California Press; 1972.
91. Sacks O. An anthropologist on Mars. New York: Vintage Books Edition; 1995.
92. Podoll K, Robinson D. Cenesthetic pain sensations illustrated by an art teacher suffering from basilar migraine. Neurol Psychiat Brain Res. 2000;8:159–64.
93. Morrison DP. Abnormal perceptual experiences in migraine. Cephalalgia. 1990;10:273–7.
94. Ro T, Farnè A, Johnson RM, Wedeen V, Chu Z, Wang ZJ, Hunter JV, Beauchamp MS. Feeling sound after a thalamic lesion. Ann Neurol. 2007;62:433–41.
95. Beauchamp MS, Ro T. Neural substrates of sound-touch synesthesia after a thalamic lesion. J Neurosci. 2008;28:13696–702.
96. Afra P, Funke M, Matsuo F. Acquired auditory synesthesia: a window to early cross-modal sensory interactions. Psychol Res Behav Manag. 2009;2:31–7. Epub. 2009 Jan 15.
97. Armel KC, Ramachandran VS. Acquired synesthesia in retinitis pigmentosa. Neurocase. 1999;5:293–6.

98. Vike J, Jabbari B, Maitland CG. Auditory-visual synesthesia. Report of a case with intact visual pathways. Arch Neurol. 1984;41:680–1.
99. Trapp BD, Nave KA. Multiple sclerosis: an immune or neurodegenerative disorders? Annu Rev Neurosci. 2008;31:247–69. https://doi.org/10.1146/annure.neuro.30.0516.094313.
100. Camichael DA, Simner J. The immune hypothesis of synesthesia. Front Hum Neurosci. 2013;7:563. https://doi.org/10.3389/fnhum.2013.00563.
101. Carruthers HR, Miller V, Tarrier N, Whorwell PJ. Synesthesia, pseudo-synethesia, and irritable bowel syndrome. Dig Dis Sci. 2012;57:1629–35.
102. Banissy MJ, Cassell JE, Fitzpatrick S, Ward J, Walsh VX, Muggleton NG. Increased positive and disorganised schizotypy in synaesthetes who experience colour from letters and tones. Cortex. 2012;48:1085–7.
103. Ellis H. Mescal: a new artificial paradise. Contemp Rev. 1898;73:130–41.
104. Luke DP, Terhune DB. The induction of synaesthesia with chemical agents: a systematic review. Front Psychol. 2013;4:753. https://doi.org/10.3389/fpsyg.2013.00753.
105. Sinke C, Halpern JH, Zedler M, Neufeld J, Emrich HM, Passie T. Genuine and drug-induced synesthesia: a comparison. Conscious Cogn. 2012;21:1419–34.
106. Elkoshi R. Is music "colorful"? A study of the effects of age and musical literacy on children's notational color expressions. Int J Edu Arts. 2004;5(2) http://ijea.asu.edu/v5n2
107. Holm J, Aaltonen A, Siirtola H. Associating colours with musical genres. J N Music Res. 2009;38(1):87–100.
108. Palmer SE, Schloss KB, Xu Z, Prado-León R. Music–color associations are mediated by emotion. Proc Natl Acad Sci. 2013;110(22):8836–41.
109. Geoffrey L. Collier affective synesthesia: extracting emotion space from simple perceptual stimuli. Motiv Emot. 1996;20(1):1–30.
110. Barbiere JM, Vidal A, Zellner DA. The color of music: correspondence through emotion. Empir Stud Arts. 2007;25(2):193–208.
111. Day SA. Regarding types of synesthesia and color music art; 2008. http://www.daysyn.com/day2008.pdf.
112. Day SA. What is synesthesia? In: Jewanski J, Day SA, Siddiq S, Haverkamp M, Reuter C, editors. Music and synesthesia. Abstracts from a conference in Vienna, scheduled for July 3–5, 2020. Wissenschaftliche Schriften der WWU Münster: Münster; 2020. p. 17–23.
113. Cutietta RA, Haggerty KJ. A comparative study of color association with music at various age levels. J Res Music Educ. 1987;35(2):78–91.
114. Marks LE. On colored-hearing synesthesia: cross-modal translations of sensory dimensions. Psychol Bull 1975. 1975;83:303–31.
115. Simner J, Ipser A, Smees R, Alvares J. Does synaesthesia age? Changes in the quality and consistency of synaesthetic associations. Neuropsychologia. 2017;106(2017):407–17.
116. Carmicha DA, Smees R, Shillcock RC, Simner J. Is there a burden attached to synaesthesia? Health screening of synaesthetes in the general population. Br J Psychol. 2019;110:530–48.
117. Odbert HS, Karwoski TF, Eckerson AB. Studies in synesthetic thinking: I. musical and verbal association of color and mood. J Gen Psychol. 1942;26:153–73.
118. Simner J, Smees R, Rinaldi LJ, Carmicha DA. Wellbeing differences in children with synaesthesia: anxiety and mood regulation. Front Biosci Elite. 2021;13:195–215.
119. Rouw R, Scholte HS. Personality and cognitive profiles of a general synesthetic trait. Neuropsychologia. 2016;88:35–48.
120. Isbilen ES, Krumhansl CL. The color of music: emotion-mediated associations to Bach's well-tempered clavier. Psychomusicol Music Mind Brain. 2016;26(2):149–61.
121. Mills CB, Boteler EH, Larcombe GK. "Seeing things in my head": a synesthete's images for music and notes. Perception. 2003;32(119):1359–76.
122. Mesz B, Sigman M, Trevisan M. A composition algorithm based on crossmodal taste-music correspondences. Front Psychol. 2012;7:636.

123. Santos N & Pulido MT. Investigation of sound-gustatory syneshesia in coffeehouse setting. Proceeding of the 4th International Conference on Internet if Things, Big Data and Security, (IoTBDS 2019). p. 294–298; 2019.
124. Crisinel AS, Spence C. As bitter as a trombone: synesthetic correspondence in non synesthetes between taste/flavors and musical notes. Atten Percept Psychophys. 2010;72(7):1994–2002.
125. Malika Auvray M, Spence C. The multisensory perception of flavor. Conscious Cogn. 2008;17:1016–31.
126. Dael N, Sierro G, Mohr C. Affect-related synesthesias: a prospective view on their existence, expression and underlying mechanisms. Front Psychol. 2013;4:754.
127. Craig AD. How do you feel – now ? The anterior insula and human awareness. Nat Rev Neurosci. 2009;10:59–70.
128. Demaree HA, Everhart DE, Youngstrom EA, Harrison DW. Brain lateralization of emotional processing: historical roots and a future incorporating dominance. Behav Cogn Neurosci Rev. 2005;4:3–20.
129. McGeoch PD, Rouw R. How everydays sounds can trigger strong emotions: ASMR, misophonia ad the feeling wellbeing. BioEssay. 2020:e2000099. https://doi.org/10.1002/bies.202000099.
130. Elkoshi R. When sound and color meet: mapping chromesthetic experiences among school children who encounter classical music. In: Forrest D, Godwin L, editors. Proceedings of the International Society for Music Education 32nd World Conference on Music Education Glasgow, Scotland 25–29 July 2016; 2016. p. 85–94.
131. Lima ESC. Developmental synesthesia, perception, and performance: challenges and new directions in music education and research. XXV Congresso da Associação Nacional de Pesquisa e Pós-Graduação em Música – Vitória – 2015;2015.
132. Bodo PPR, Schiavoni LF. Web Sonification with synesthesia tools. Música Hodie, Goiânia. 2018;18(1):74–91.
133. Konanchuk SV, Grigorenko AY. Synesthesia and virtual communication: aesthetic approaches to research. *IEEE communication strategies in digital society workshop (ComSDS)*; 2018. p. 29–32. https://doi.org/10.1109/COMSDS.2018.8354959.

Music and Creativity: The Auditory Mirror System as a Link between Emotions and Musical Cognition

Barbara Colombo

3.1 Introduction

Mirror neurons (part of the mirror neuron system—MNS) diverge from motor and sensory neurons due to the fact that they become active both with the performance of an action and with the observation of another performing the action [1, 2]. In humans, the MNS helps understand others' actions and the intentions behind them [3], and it has also been suggested that it has an important role in mediating empathy [4]. Recent research suggested that the mirror system is not activated only by visual stimuli and involves the auditory system as well. This has been proven by the fact that a group of audiovisual neurons in the ventral premotor F5 area seems to be able to discriminate between different actions with extremely high accuracy when only seen or only heard [5, 6]. Following up on this line of research, more evidence highlighted how there are auditory mirror neurons in humans that fire in response to the sounds of actions that individuals are capable of performing [7]. Since the representation of sensory and motor information in the human brain is integrated at many levels, seeing or hearing action-related stimuli automatically cues the movements required to respond to or produce them, in order to guide the perception of musical stimuli [8]. Studying this fascinating relationship, the role that the MNS might play in facilitating or mediating the understanding of music has been investigated [9]. Focusing on the role of the MNS in professional musicians when they were listening to music, a recent study [10] found that auditory mirror activation only occurred when listening to a passage from a song that participants were taught to play, and did not happen when listening to a passage of an unfamiliar song. The researchers explained this finding by hypothesizing that only sounds within our motor repertoire will activate the auditory MNS; hence, the musicians did not respond to songs they had not been

B. Colombo (✉)
Behavioral Neuroscience Lab—Champlain College, Burlington, VT, USA
e-mail: bcolombo@champlain.edu

taught due to their unfamiliarity with them [10]. Another study [11] discussed how musicians could have a good understanding of the piano without either motor or auditory stimuli, because of their deeper knowledge of both auditory and motor components of piano playing. This reading stands on the assumption that professional musicians would show more MNS activity in response to both familiar and new music than other individuals. Other research data supports the fact that mirror neuron activation is modulated by musical expertise and that MNS activation in musicians. Moreover, this specific activation could be linked to a unconscious form of imagery, which leads to imagining themselves playing the piece that they are hearing. This would also explain why the activation is presumably stronger when musicians listen to music performed on their main instrument [12].

As we briefly mentioned above, it suggests that the MNS serves as a link and common neural substrate when processing motor information and emotional as well as some other high-level cognitive information (for example, some form of learning) [13]. The link with emotions has been proven to be very interesting for researchers [14, 15] who, in the light of the possible existence of a specific auditory MNS, explored its role in helping to discriminate emotions. Results from this line of research highlighted how, when listening to different vocalizations, distinct functional subsystems within the auditory–motor mirror network respond differently to their emotional valence and arousal properties. For example, it has been reported [14] how listening to nonverbal vocalizations (which can be compared to some extent to musical sounds in the sense that they are nonverbal) can lead to an automatic preparation of specific responsive gestures. This happens fastest and more frequently for positive-valence and high-arousal emotions. If the connection between these results and possible similar responses activated specifically by music might seem logic, the specific role and the specific response played by the MNS when musicians listen to music featuring their main instrument has only been partially explored [16].

The study discussed in this chapter aimed to provide some additional evidence on if and how the level of activation of the MNS affects either the emotion response triggered by the music or the evaluation of musical creativity in a sample of professional musicians.

We decided to add a specific focus on creativity for two main, research-based, reasons.

First, the relationships between creativity and empathy (and hence creativity and the MNS) is supported by the notion that creativity is linked to and supported by social aspects [17]. For this reason, an individual will be more creative when connected to other people's minds and feelings [18]: as we discussed above, this aspect is also linked to and promoted by the activation of the MNS. Creative activities, such as painting [19] or creative dancing [20], have been shown to be a useful resource to promote empathy and other related social skills [21]. Empathy not only can be increased by creative activities, but it also affects how individuals perceive and emotionally respond to performing arts, like music [22]. It is not surprising that these findings can be applied to music since not only music has been defined as a type of creative thinking [23], but since music listening and music performing are generally social activities, just listening to music in humans has been reported to involve empathic responses [22, 24–26].

Second, something that has been clearly established within the academic field is the direct relationship between empathy and the MNS [27]. This relationship includes a positive correlation between motor and facial mimicry and empathy scores [28], affecting both visual and auditory pathways in the MNS, as well as a positive correlation between perspective taking empathy scale scores and the activation of the mirror system [7]. Moreover, the relationship between cognitive empathy and emotional states that allows us to understand others' emotions by referring to our own experience [29] can be seen as similar to the process that allows individuals (especially musicians, as discussed above) to refer back to their own motor experience to better "frame" and understand a sound produced by another individual [7, 10].

Starting from this background, in this chapter we present and discuss some data aimed at exploring the involvement of the auditory MNS in professional musicians when they listen to music. In our study, we used transcranial direct current stimulation (tDCS) to inhibit the activation of the MNS, and the measured professional musicians' emotional and cognitive responses to a new piece of music involving the instrument they play. To be more specific, we investigated how cathodal tDCS stimulation of musicians' brain area associated with the MNS would affect their judgment of how creative the music was as well as their emotional response to it.

Since cathodal tDCS has been proven to reduce the activation of the targeted area, we expected that participants who received cathodal tDCS would rate the music as less creative when compared to participants in the sham condition, given the fact that their auditory MNS would be impaired. Similarly, we hypothesized that cathodal tDCS would impact self-reported emotional reactions to music, by way of reducing the intensity of reported emotions.

3.2 Methods

The study has been reviewed and approved by Champlain College IRB.

3.2.1 Sample

Forty young musicians (age range: 18–22, mean = 19.80; SD = 1.56; $z = 15$) joined the study and were randomly assigned either to the experimental group (cathodal stimulation) or to the control group (sham stimulation).

Participants were screened before being invited to join the experiment by checking that their principal instrument would be either piano, violin, or cello (the instruments played in the piece of music used during our experiment). We also verified that they would practice a minimum of 4 h a day and have performed in public in a professional setting at least 5 times. Of the recruited participants, 16 were piano players, 14 were violinists, and 10 cellists.

3.2.2 Procedure and Instruments

Procedure is described in Fig. 3.1.

Fig. 3.1 Procedure

3.2.2.1 tDCS Equipment

In this study, we used 1300A 1 × 1 transcranial direct current low-intensity stimulator by Soterix Medical to deliver brain stimulation to our participants. We used two 5 × 5 cm rubber electrodes enveloped in saline-soaked sponges covered with conductive gel. For the experimental conditions (cathodal), the stimulation was set at 1.5 mA for 20 min. In the control (sham) condition, the equipment started the stimulation normally and ramped up to the target intensity of 1.5 mA; it decreased to 0 mA after 5 s. This gave participants the impression of receiving stimulation, when in reality the stimulation lasted only 5 s, thus having no actual effect on brain functions. For the experimental condition, the electrodes were placed on the left ventral premotor cortex using the 10–20 system (F5 location). The anodal electrode was placed on the upper right forearm. The same montage was used for the sham condition.

3.2.2.2 Geneva Emotional Scale (GEW)

The Geneva Emotion Wheel (GEW) [30, 31] measures emotional reactions to objects, events, and situations. Participants are asked to indicate the emotion(s) they experienced by choosing intensities for a single emotion or a blend of several emotions out of 20 different options. The emotions are arranged in a wheel shape, with the axes being defined by two major dimensions of emotional experience: high vs. low control and positive vs. negative valence. Five degrees of intensity are being proposed, represented by circles of different sizes. In addition, "None" (no emotion felt) and "Other" (different emotion felt) options are provided.

3.2.2.3 Creativity Evaluation

We asked participants to rate specific factors that have been reported in the literature to be associated with creativity: interest [32–34], innovation [35–37], and excitement [38, 39]. Participants were asked to rate the creativity of the musical piece by rating how interesting, innovative, and exciting the piece was on a 9-points Likert scale. To be more specific, participants were told: "You are now asked to evaluate the creativity of the piece you just listened to. How interesting/innovating/exciting you think it is?"

3.2.2.4 Music

Dreaming Cities is a five-movement piano trio (violin, cello, piano) by Damon Ferrante (see Fig. 3.2). In this experiment, participants listened to the third movement. The third movement is a slow movement whose material is a variation of the musical theme that occurs at the beginning of the work. The third movement's sparce,

Fig. 3.2 Excerpt from "Dreaming Cities," reproduced with permission from the author

lyrical texture highlights the characteristic musical voices of each instrument. It was not written with a specific emotional tone in mind, but, rather, focusing on the slow, melodic interplay of the instruments. This piece of music was not familiar to any participant (a familiarity check was performed at the end of the experiment).

3.3   Results

To explore the effects of the brain stimulation on emotional reaction as well as creative evaluation of the musical piece, we ran a GLM MANOVA, using the condition as an independent variable and the three creative evaluation scales (interest, innovation, and excitement) and self-report of emotional response (categorized into two variables: sum of positive valence emotions and sum of negative valence emotions) as dependent variables. Figures 3.3 and 3.4 show the mean scores for creative evaluations and emotional responses for the two tDCS conditions.

The test of between-subject effects returned a significant main effect of stimulation condition on the evaluation of the creativity of the piece. Two of the considered dimensions were significantly affected: how innovative the piece was ($F_{1;34} = 45.76, p < 0.001, \eta^2 = 0.57$) and how exciting it was ($F_{1;34} = 53.73, p < 0.001, \eta^2 = 0.61$). In both cases, cathodal stimulation decreased the reported perception of creativity.

Focusing on the self-report emotional response to the piece, cathodal stimulation significantly affected emotions with negative valence ($F_{1;34} = 17.93, p < 0.001, \eta^2 = 0.34$). Cathodal stimulation decreased the intensity of negative emotions reported by participants.

When analyzing the effect of tDCS on specific positive emotional responses that can be affected by listening to music (namely, interest and admiration), we see how cathodal stimulation also reduced them (see Fig. 3.5), with the difference being significant for interest ($F_{1;40} = 7.60, p = 0.009, \eta^2 = 0.318$). Adding age as a covariate, it had a significant effect in moderating the relationship between the tDCS condition and the emotional response, as can be seen in Fig. 3.6, with admiration being affected substantially in older musicians and the effects being less pronounced overall in younger musicians.

Fig. 3.3 Mean scores of creative evaluation by condition

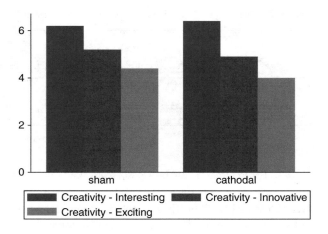

Fig. 3.4 Mean scores of emotional valence by condition

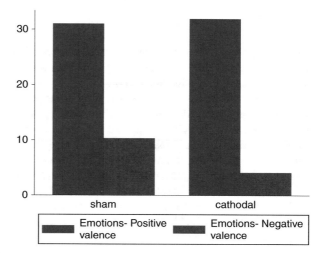

Fig. 3.5 Mean scores of interest and admiration by condition

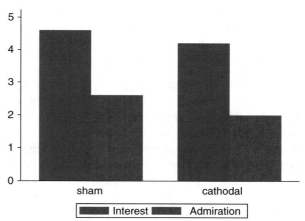

Fig. 3.6 Mean scores of interest and admiration by condition and by age groups

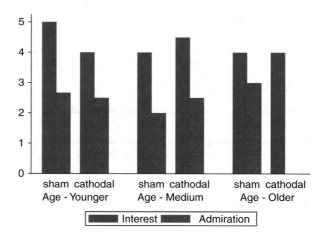

3.4   Discussion and Conclusions

In this chapter, we presented some data on the role of the auditory mirror system in influencing the evaluation of creativity as well as the emotional reactions of professional musicians when listening to music.

We were especially interested in exploring the effect of cathodal stimulation in reducing the perceived creativity of the new piece of music. This interest was inspired by research evidence that reports how the auditory MNS plays a role in musicians' response to music [10, 11] and also that the MNS's role is linked to processing not only motor information but also emotional and other higher-level cognitive information, like, for example, creativity [13]. As we discussed in the introduction, empathy, affected by the MNS [4, 7, 29], plays a role in affecting how individuals evaluate performing arts, including music [22].

The results described in this chapter provide interesting new insights. After cathodal stimulation, musicians tended to perceive music as less innovative and exciting when compared to musicians who underwent sham stimulation. On the other hand, their evaluation of the level of interest was not significantly affected by the stimulation. Our data seem to confirm the role of the MNS in evaluating the creativity of a music piece, but the role seems to be rather specific. Both the cognitive evaluation of the creative process (the innovation of the piece) and the emotional reaction to it (excitement) appear to be influenced by the activation of the MNS. When the activation is lowered by cathodal stimulation, the piece is perceived as less innovative and less exciting. On the other end, how interesting the piece is appears to be examined through a different circuit. We might hypothesize that this evaluation can be related to individual differences and hence not being directly affected by the modulation of the MNS. This reading is supported by research data stating that music preference is significantly influenced by a combination of the individuals' perception of the cognitive, emotional, and cultural functions of music, together with the physiological arousal and familiarity [40]. This reading is also supported by the fact that, when taking age into consideration, the effect of the tDCS in influencing how mush participants admired the piece or found it interesting varied considerably by age group, implying, maybe, an important role of expertise in mediating the effect of the brain stimulation. Further research might include the evaluation of these variables into a tDCS design similar to the one presented in this chapter. Something else that would be interesting to consider, and that wasn't controlled for in this study, is participants' level of attention [41, 42].

We also examined the effect of tDCS on participants' reported emotions after listening to the music. This interest was inspired by the evidence supporting the fact that the auditory MNS plays a significant role in responding to auditory stimuli with emotional valance [14, 15], like, for example, music. We found and reported a significant effect of cathodal stimulation on participants' self-report of emotions with a negative valence: after cathodal stimulation, musicians involved in the study reported less emotion with a negative valence. We can explain this by referring to the specific music we were using for our study. Even if the movement that we used was not written with a specific emotional tone, it has a slow tempo and it is mainly written in the tonality of D minor. Minor keys and lower tempos tend to be associated with more negative emotions like sadness [43], so the effect of neuromodulation might have been more pronounced for emotions that are linked to sadness. Also, fMRI data suggest that familiarity seems to play an important role in making the listeners emotionally engaged with music [44], and our piece was unfamiliar to all our participants.

Even if the data that we discussed here cannot be considered final, and more evidence is needed, it sheds some light on the role of the auditory MNS in evaluating specific aspects of musical creativity (innovation and excitement) and in influencing, to some extent, the emotional response to the same music, by offering some more evidence that can help clarify the role of the auditory MNS in evaluating music.

These data also offer some food for thought regarding practical applications. They suggest concrete possibilities for new uses of music to promote creativity as well as social skill in different educational settings. To be more specific, the fact that music can affect both creativity and empathy could be used to build specific interventions aimed at working with youth with autism spectrum disorders [45] but could also be used to inform assessment in music composition [46].

Future studies should include anodal stimulation to compare the effects of a different activation of the MNS; they could also use music characterized by different tempo and/or keys and take musicians' age and expertise into consideration by using a larger sample.

References

1. Kilner JM, Lemon RN. What we know currently about mirror neurons. Curr Biol. 2013;23(23):R1057–62.
2. Rizzolatti G. The mirror neuron system and its function in humans. Anat Embryol. 2005;210(5–6):419–21.
3. Cattaneo L, Rizzolatti G. The mirror neuron system. Arch Neurol. 2009;66(5):557–60.
4. Baird AD, Scheffer IE, Wilson SJ. Mirror neuron system involvement in empathy: a critical look at the evidence. Soc Neurosci. 2011;6(4):327–35.
5. Keysers C, Kohler E, Umiltà MA, Nanetti L, Fogassi L, Gallese V. Audiovisual mirror neurons and action recognition. Exp Brain Res. 2003;153(4):628–36.
6. Kohler E, Keysers C, Umilta MA, Fogassi L, Gallese V, Rizzolatti G. Hearing sounds, understanding actions: action representation in mirror neurons. Science. 2002;297(5582):846–8.
7. Gazzola V, Aziz-Zadeh L, Keysers C. Empathy and the somatotopic auditory mirror system in humans. Curr Biol. 2006;16(18):1824–9.
8. Stephan MA, Lega C, Penhune VB. Auditory prediction cues motor preparation in the absence of movements. NeuroImage. 2018;174:288–96.
9. Jiang J, Liu F, Zhou L, Jiang C. The neural basis for understanding imitation-induced musical meaning: the role of the human mirror system. Behav Brain Res. 2019;359:362–9.
10. Lahav A, Saltzman E, Schlaug G. Action representation of sound: audiomotor recognition network while listening to newly acquired actions. J Neurosci. 2007;27(2):308–14.
11. Bangert M, Peschel T, Schlaug G, Rotte M, Drescher D, Hinrichs H, Heinze H-J, Altenmüller E. Shared networks for auditory and motor processing in professional pianists: evidence from fMRI conjunction. NeuroImage. 2006;30(3):917–26.

12. Hou J, Rajmohan R, Fang D, Kashfi K, Al-Khalil K, Yang J, Westney W, Grund CM, O'Boyle MW. Mirror neuron activation of musicians and non-musicians in response to motion captured piano performances. Brain Cogn. 2017;115:47–55.
13. Ramachandra V, Depalma N, Lisiewski S. The role of mirror neurons in processing vocal emotions: evidence from psychophysiological data. Int J Neurosci. 2009;119(5):681–91.
14. Warren JE, Sauter DA, Eisner F, Wiland J, Dresner MA, Wise RJ, Rosen S, Scott SK. Positive emotions preferentially engage an auditory–motor "mirror" system. J Neurosci. 2006;26(50):13067–75.
15. Banissy MJ, Sauter DA, Ward J, Warren JE, Walsh V, Scott SK. Suppressing sensorimotor activity modulates the discrimination of auditory emotions but not speaker identity. J Neurosci. 2010;30(41):13552–7.
16. Colombo B, Anctil R, Balzarotti S, Biassoni F, Antonietti A. The role of the mirror system in influencing musicians' evaluation of musical creativity. A tDCS study. Front Neurosci. 2021;15:298.
17. Glaveanu V, Lubart T, Bonnardel N, Botella M, De Biaisi P-M, Desainte-Catherine M, Georgsdottir A, Guillou K, Kurtag G, Mouchiroud C. Creativity as action: findings from five creative domains. Front Psychol. 2013;4:176.
18. Form S, Kaernbach C. More is not always better: the differentiated influence of empathy on different magnitudes of creativity. Eur J Psychol. 2018;14(1):54.
19. Gerry LJ. Paint with me: stimulating creativity and empathy while painting with a painter in virtual reality. IEEE Trans Vis Comput Graph. 2017;23(4):1418–26.
20. Batson G. Chapter 6: Sharing creativity through the mirror neuron system: embodied simulation through dance. In: Creativity and entrepreneurship: changing currents in education and public life. Cheltenham: Edward Elgar Publishing Ltd; 2013. p. 66.
21. Morizio LJ. Creating compassion: harnessing creativity for empathy development. Boston: University of Massachusetts; 2021.
22. Wöllner C. Is empathy related to the perception of emotional expression in music? A multimodal time-series analysis. Psychol Aesthet Creat Arts. 2012;6(3):214.
23. Antonietti A, Colombo B. Musical thinking as a kind of creative thinking. In: Creativity research: an inter-disciplinary and multi-disciplinary research handbook. Milton Park: Routledge; 2014. p. 233.
24. Balteş FR, Miu AC. Emotions during live music performance: links with individual differences in empathy, visual imagery, and mood. Psychomusicol Music Mind Brain. 2014;24(1):58.
25. Sittler MC, Cooper AJ, Montag C. Is empathy involved in our emotional response to music? The role of the PRL gene, empathy, and arousal in response to happy and sad music. Psychomusicol Music Mind Brain. 2019;29(1):10.
26. Cross I, Laurence F, Rabinowitch T-C. Empathy and creativity in group musical practices: towards a concept of empathic creativity. In: The Oxford handbook of music education, vol. 2. Oxford, Oxford University Press; 2012.
27. Bekkali S, Youssef GJ, Donaldson PH, Albein-Urios N, Hyde C, Enticott PG. Is the putative mirror neuron system associated with empathy? A systematic review and meta-analysis. Neuropsychol Rev. 2020;1:1–44.
28. Sonnby-Borgström M, Jönsson P, Svensson O. Emotional empathy as related to mimicry reactions at different levels of information processing. J Nonverbal Behav. 2003;27(1):3–23.
29. Schnell K, Bluschke S, Konradt B, Walter H. Functional relations of empathy and mentalizing: an fMRI study on the neural basis of cognitive empathy. NeuroImage. 2011;54(2):1743–54.
30. Scherer KR. What are emotions? And how can they be measured? Soc Sci Inf. 2005;44(4):695–729.
31. Scherer KR, Shuman V, Fontaine J, Soriano Salinas C. The GRID meets the wheel: assessing emotional feeling via self-report. In: Components of emotional meaning: a sourcebook. Oxford: Oxford University Press; 2013.
32. Li H, Li F, Chen T. A motivational–cognitive model of creativity and the role of autonomy. J Bus Res. 2018;92:179–88.

33. Moreira IX, da Costa A, Belo L, dos Santos GA, Savio R. Impact of creativity and interest in learning on student achievement Instituto superior Cristal students. J Innovat Stud Char Edu. 2020;4(1):70–8.
34. Fürst G, Grin F. A comprehensive method for the measurement of everyday creativity. Think Skills Creat. 2018;28:84–97.
35. Acar S, Burnett C, Cabra JF. Ingredients of creativity: originality and more. Creat Res J. 2017;29(2):133–44.
36. Lee A, Legood A, Hughes D, Tian AW, Newman A, Knight C. Leadership, creativity and innovation: a meta-analytic review. Eur J Work Organ Psy. 2020;29(1):1–35.
37. Rietzschel EF, Ritter SM. Moving from creativity to innovation. In: Individual creativity in the workplace. Amsterdam: Elsevier; 2018. p. 3–34.
38. Fink A, Reim T, Benedek M, Grabner RH. The effects of a verbal and a figural creativity training on different facets of creative potential. J Creat Behav. 2020;54(3):676–85.
39. Paulus PB, Nijstad BA. The Oxford handbook of group creativity and innovation. Oxford: Oxford Library of Psychology; 2019.
40. Schäfer T, Sedlmeier P. What makes us like music? Determinants of music preference. Psychol Aesthet Creat Arts. 2010;4(4):223.
41. Li H, Duan H, Zheng Y, Wang Q, Wang Y. A CTR prediction model based on user interest via attention mechanism. Appl Intell. 2020;50(4):1192–203.
42. Peters C, Pelachaud C, Bevacqua E, Mancini M, Poggi IA. Model of attention and interest using gaze behavior. In: International workshop on intelligent virtual agents. Berlin: Springer; 2005. p. 229–40.
43. Webster GD, Weir CG. Emotional responses to music: interactive effects of mode, texture, and tempo. Motiv Emot. 2005;29(1):19–39.
44. Pereira CS, Teixeira J, Figueiredo P, Xavier J, Castro SL, Brattico E. Music and emotions in the brain: familiarity matters. PLoS One. 2011;6(11):e27241.
45. Forti S, Colombo B, Clark J, Bonfanti A, Molteni S, Crippa A, Antonietti A, Molteni M. Soundbeam imitation intervention: training children with autism to imitate meaningless body gestures through music. Adv Autism. 2020;6:227–40.
46. Deutsch D. Authentic assessment in music composition: feedback that facilitates creativity. Music Educ J. 2016;102(3):53–9.

A Short History of Rhythm

4

David Hildebrandt

4.1 Rhythms around us

Beat beat beat beat … the fundamental pulse of all life. Pulse is essential to life. Whether it's from the heart or the breath. Pulse is inside us, but also all around us. In addition to life itself, pulse gives us regularity and periodicity, which is the mother of any kind of rhythmic evolution. No matter what we do as humans, it's influenced by the phenomenon of rhythm. Whether we breathe, walk, or talk, rhythm is an integral part of life. For humans the most fundamental rhythm, however, must be the rhythm of cosmos, as it is the cause of rhythm here on Earth. The rotation of the Earth gives us the alternation of light and dark. Day and night. This is one of the most striking examples of global rhythms. We group days into months by the orbital period of the Moon around Earth, and into years by the Earth's orbit around the Sun. These cycles, or rhythms, have a tremendous impact on earthly life, affecting virtually everything including human, plant, and animal life. The rhythm of day and night causes the periodicity of sleeping and awakening, which applies for all creatures on planet Earth. As humans, we have a need for rhythm. Regularity and repeating. Seconds are grouped in minutes, minutes are grouped in hours, and hours are grouped in days and weeks. Daily routines are repeated week after week. Month after month. Year after year. A human life is one long cycle of pure rhythm.

4.2 Connecting Through Pulse

Even though we are dealing with rhythm every day, most people mistakenly think that they have a weak sense of rhythm when it comes to music. But rhythm is such an essential part of life that most of us are actually pretty good at it, without

D. Hildebrandt (✉)
The Royal Danish Academy of Music Percussion, Copenhagen, Denmark

B. Colombo (ed.), *The Musical Neurons*, Neurocultural Health and Wellbeing,
https://doi.org/10.1007/978-3-031-08132-3_4

knowing. Think of this phenomenon: A show is over and it was great. Therefore, we must pay tribute to the artists by clapping. At first it's a mess. Everyone is clapping at their own pace and it creates a so-called white noise. But then the audience is gradually finding a common beat. A little slower than before, but rhythmic and persistent. How is this possible? How are hundreds of people locking into the same tempo as one big organism? Well, first of all, it comes from a collective wish to express cohesion with each other and recognition of the artists on stage. If we synchronize the applause, it actually becomes remarkably louder than the white noise produced by several individual tempos. This makes it also a social statement. If we as a crowd want to communicate that the show was a blast, it becomes more significant in the ears of the performers on stage when we synchronize our claps. And as humans we are extremely good at synchronizing. Although in theory it is more than difficult to turn white noise into one single rhythm, it happens easily when we agree on it. Basically, we are affected by the clapping of others so fast that we don't even notice. If the guy next to you at the audience is clapping at a slightly slower tempo, you may adjust to his tempo without noticing. Now the two of you are interlocked. And if the two more people join in, you have a unison group. This group may lock in with another group etc., and suddenly 1000 people may agree to clap at exactly the same speed! It is simply quite natural for humans to adjust to surroundings by using pulse as a sophisticated socio-cognitive form of communication.

4.3   Meter

Although clapping together is a rather primitive form of rhythm, it is nonetheless the concept of steadiness; that's the fundament in more sophisticated rhythms. In music, a steady pulse, or simply a tempo, is measured in beats per minute (bpm). A steady pulse alone is so simple that we may even not regard it as rhythm, but just as a series of beats equally spaced in time: $1 - 1 - 1 - 1 - 1 - 1 - 1 - 1$... An endless series of undifferentiated beats. But how does such an elemental thing as pulse become so sophisticated an art as a musical rhythm? If we systematically mark a series of beats into regular groups of (e.g.) 4 and accentuate every first beat of each group, a pattern will emerge: $\boldsymbol{1} - 1 - 1 - 1 - \boldsymbol{1} - 1 - 1 - 1 - \boldsymbol{1} - 1 - 1 - 1$. Because the beats appear in groups of 4, they could also be counted: $1 - 2 - 3 - 4 - 1 - 2 - 3 - 4$. Suddenly, pulse has become meter. We have now taken the first step toward rhythmic art, by measuring and controlling pure pulse and putting it into bars, as the strong and accentuated beat defines the first beat of each bar. The downbeat. Meter (the doctrine of bars and beats) is the groundwork of rhythm. Still meter alone is quite far away from rhythm as a form of art. But what happens if we hear two different tempos at the same time? Like 40 and 30 bpm played simultaneously. One may think that our ears will just perceive two different speeds, which is right and wrong at the same time. When played simultaneously, the two tempos merge into *one* single rhythm consisting of strong and weak beats. A real rhythm is created – a rhythm that is much more than just a monotonous series of beats, but a playful phrase with twists and surprises.

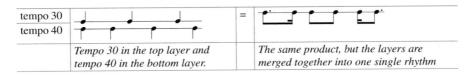

tempo 30					=					
tempo 40										
Tempo 30 in the top layer and tempo 40 in the bottom layer.						*The same product, but the layers are merged together into one single rhythm*				

The product of the two tempo layers – the composite – is also known as a polyrhythm. Polyrhythms exist in all cultures and are one of the oldest types of rhythms that we know of. Simple ratios are used to describe polyrhythms. Like 4:3 (tempo 40 and 30) – pronounced four over/against three. In 4:3, 4 refers to the number of units in the fastest layer, and 3 refers to the number of beats in the bar (the meter).

An interesting aspect about this is that a polyrhythm can be heard and perceived in multiple ways. When the human ear is exposed to more than one tempo (a polyrhythm), it will automatically select one of them as the main tempo. If tempos 40 and 30 are played at the same time, and the ear chooses tempo 40 as the main tempo (3:4), the composite of the two tempos will sound completely different than the rhythm that occurs if tempo 30 is selected as the main tempo (4:3). Even though the rhythmic products of 3:4 and 4:3 technically speaking are exactly the same, they are extremely different when it comes to sound and feel. This phenomenon can be seen as rhythmic ambiguity and is a sonic counterpart to the famous picture with the faces and the vases.

4.4   Speech

Language can be used to illustrate the difference between rhythms that technically speaking are the same – like the composites of tempo 40 and 30. In language, we have a natural way of phrasing strong and weak beats. And that is what rhythm basically is – a combination of strong and weak beats. Strong beats are directly connected to a pulse, and weak beats live around this main pulse. See the two sentences below. Both consist of 6 syllables. But the first has 3 strong beats and 3 weak. The second has 4 strong and 2 weak.

Tempo 30 as the primary tempo	Tempo 40 as the primary tempo
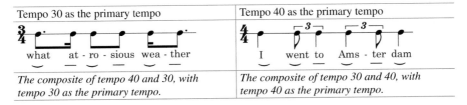	
The composite of tempo 40 and 30, with tempo 30 as the primary tempo.	*The composite of tempo 30 and 40, with tempo 40 as the primary tempo.*

So even though the rhythmic product of the two examples above technically is exactly the same, as the time intervals between each syllable are identical in both bars, language shows us how different they feel depending on which of the two tempos our ears choose as the primary.

Through the language we are all in contact with rhythm on a daily basis. And quite sophisticated rhythms actually. Speech becomes rhythmical not only because it consists of sounds and pauses, but also because every word that contains more than one syllable consists of a combination of strong and weak syllables like the two examples above. Furthermore, we have the interaction between vowels and consonants. And of course the aspect of timing and phrasing. Let's have a look at the famous line from Shakespeare's Hamlet:

To be, or not to be, that is the question

This phrase takes its departure from the standard iambic parameter (weak – strong), but from here the relations between strong and weak syllables are varied a great deal.

To be, or not to be - that is the question

In order to exhibit the similarities in language and rhythm, I found four famous recordings of the scene and transcribed each interpretation into rhythm.

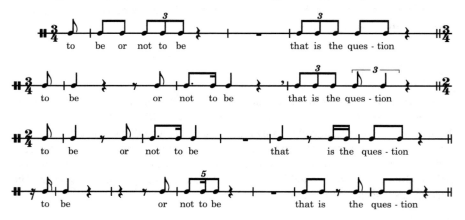

There are endless ways to phrase the line that all would emphasize the meaning of the words differently. But as you can see above, one thing applies to all examples; all strong syllables lie directly on a pulse-beat, and all weak syllables live around the beats. This illustrates the hierarchy in the rhythms clearly. The strong beats define the fundamental structure of the line, while the weak beats add the life, twists, syncopations, and surprises. There are of course exceptions to the idea of strong and weak beats. Offbeats, for example, are often quite accentuated, even though they are "off beat."

Furthermore, it must be mentioned that none of the four examples above were performed with a strict metronomic precision. If read in strict musical time, the output would seem somewhat forced and mechanical. However, this fact doesn't make the comparison between language and music less relevant, as we in many musical genres often tend to avoid a mechanical and strict approach to rhythm and phrasing, in order to make music appear more free, alive, and "speaking."

4.5   Structure

As you can see from the examples above, different meters are used to define the line. Both 3/4 and 2/4 are used here. These meters primarily tell us something about the spacing and distance between the downbeats; the strongest and most stressed words in the phrase. Here, and in most other cases, meter on its own is a pretty doll subject. Whether music is in 2/4, 4/4, or 9/8, it's not fascinating information. It's more a technicality, and it doesn't contain much musical value alone. The subject, however, becomes quite fascinating when we look at the larger aspects of music. By tracing how meters grow into rhythmic phrases, and how these phrases build into larger periods and again into whole movements, we can understand the essence of musical structure much better.

In order to understand this process of rhythmic growth, we must first understand the principle of doubleness; the concept of two. Almost all music is somehow built on a double concept. Why do we need this duplicity so badly? Earlier I described how all rhythm derives from the steady pulse. Pulse is essential to man since our heart pumps blood around in our system and keeps us alive. But there are two phases to every heartbeat. Expand and contract. Physical life is double. We inhale – we exhale. There are no third steps in this process. This is the symmetry our bodies are based on. And (as Leonard Bernstein puts it), *as two-legged creatures we walk left, right, left, and right into the heart of music*. It somehow feels natural that most music has a double meter – two beats per bar. Or some multiples of two beats per bar like 4/4, 6/4, or 8/4. Our biological need for duplicity is so great that we even tend to group bars dualistically.

Let's have a look at the children's song: *London Bridge is Falling Down*. The entire melodic phrase looks like this:

If we listen to the melody in the first bar, we will only hear a group of notes with no musical meaning on their own. This bar obviously has to be paired with a complementary bar, to achieve even the most elementary musical sense.

Still these two bars alone feel quite unsatisfying.	In order to arrive at a single phrase, we need another pair of bars to balance the first two.

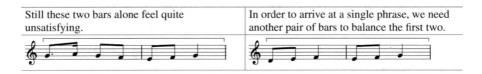

Putting together these two pairs of bars, we've now got a phrase consisting of four bars:

Standing alone, the four bars still feel incomplete. Therefore, we supply these four bars with another matching four-bar phrase, to create symmetry and complete the sentence:

The phrase, consisting of 2 × 2 × 2 bars, can easily stand alone with a valuable musical meaning (Leonard Bernstein refers to this kind of rounded phrase as a *sentence*). The example above is of course rather short, but if we analyzed a song or a piece consisting of 16 bars, we would discover that an 8-bar group most likely would be paired with another 8 bars. The second group of 8 bars would balance up the first 8 bars creating a total of 16 bars (2 × 2 × 2 × 2) and so on. The geometric progression is clear, and it all derives from the metrical unit of two.

Now, many of you might be thinking: What about all the music with a meter consisting of 3 beats per measure? Like the theme from Bach's Goldberg Variations. Or all the waltzes by Strauss. Or Beatles' evergreen Norwegian Wood. Isn't music in 3 almost as fundamental as music in 2? It is. But just almost. The number 3 is not bounded in our biological nature. We just don't breathe in 3. Nor does the heart beat in 3. Yet music in 3 feels just as valuable as music in 2. The answer to this paradox lies partly in the concept of contrast. As we know from the melodic phrase above, the form is somehow reducible to 2. But let's have a closer look at 2. According to the theory of duplication, 2 is created by 1, which therefore must be the almighty mother of 2.

$$1+1=2$$

Now, if we go the Fibonacci path and pair 1 with itself, we get 2. And if we again pair 2 with its creator, 1, we get the golden number of 3.

$$1+1=2, 1+2=3$$

3 is a contrast to 2. It's the first and finest exception to our natural instinct of left – right. 1, 2, and 3 are without a doubt the most important numbers in rhythm

and in music generally speaking. One might say that a meter of 3 beats per bar is an intellectual or constructed meter, as it is primarily an unphysical concept.

We've now reached an important point in our understanding of rhythm. Having investigated what pulse is and how it becomes a meter of either a double or triple kind, we are now in possession of all the elements needed to understand any rhythm. All rhythms are in one way or another a result of the interaction of physical 2 and intellectual 3. It could be 2 + 2 (resulting in a meter of 4/4). Or 3 + 2 or 3 + 2 + 2 or 3 and 2 simultaneously that would create a polyrhythm. Somehow all rhythms are reducible to the cells of 2 and/or 3. Even in music that is composed in a meter of 3, the concept of 2 lies just beyond the surface. Just take a look at the lullaby below: *Rock-a-Bye-Baby*. Even though the meter is in 3, the line is subject to the same rules of duplexity and multiplication as described in the previous example.

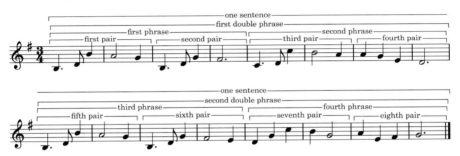

Even though this example is in 3, it turns out to be just as much a slave to doubleness as the previous example in 2. This is most likely because our biological 2 lies just beneath the surface. The 3/4 bar is not just *one two three, one two three,* etc., but more **left** *two three,* **right** *two three,* **left** *two three,* **right** *two three.* And so the overall structure in triple metered music turns out to be just as much double as a polka.

In music, our biological demand for doubleness also becomes an aesthetic demand, and that is why symmetry and balance are keywords when it comes to musical structure. Beats combine symmetrically into bars, bars combine symmetrically into phrases, phrases combine symmetrically into larger formal sections, and sections combine symmetrically into whole movements. This has been the spine in western music up to the twentieth century (but also in many other art forms). In the beginning of the twentieth century, many composers started to play around with 5 and 7 bar periods, but these unsymmetrical periods were mostly paired with other groups of 5 or 7 bars and therefore still bound to duplexity.

However, in modern western music, a rhythmic "revolution" led by composers such as Aaron Copland and Igor Stravinsky finally turned the symmetrical tyranny upside down, creating music of wild, unpredictable, and irregular structure and rhythm. Music with contrasting meters and rhythms combined with unsymmetrical bar periods became a new reality. This kind of savage and wild rhythm was never heard before in western classical music. The rhythms seemed somehow new, somehow raw, and somehow authentic. Maybe the concept of rhythm was just taken back to origins? The context in which these rhythms were used was brand new, but maybe the rhythms themselves were not so new after all?

4.6   Origins

The wild rhythms (and musical material in general) that Igor Stravinsky created in his revolutionary *Le Sacre du Printemps* (premiered 1913) were extremely advanced according to Western standards. The work was so controversial at the time that the premiere ended in tumult and fistfights. Although the rhythms sounded modern, they also had a primitive and ritual quality that added a whole new dimension to "classical" music. As a composer, Stravinsky was a product of the western "cultivated" tradition, and in that sense, his music was "art music." But ironically, the rhythms he used were "borrowed" from ancient African music, where aesthetics is a secondary question and is not a goal in itself. The European notion "art for art's sake" doesn't make sense in the African tradition. Traditional African music exists to serve something more than itself. It has some kind of a purpose beyond its own existence – whether ritual, ceremonial, or simply for entertainment.

In western music, there is a radical division between the *active* musicians and the *passive* audience that just doesn't exist in the same way in African traditional music. African music is a collective matter, and every member of a society would somehow participate, although not necessarily in the same degree. The concept of going out to "listening to music" is not a thing. Making music together is. Some play, others sing or dance, but everyone participates one way or another. The social aspect is a fully integrated part of traditional African music.

Even though African music in many ways is rather understandable and "easy to listen to" for most people around the globe, it also has a high level of rhythmic complexity. It has a circular character with a less "mathematical" approach to meter than European music. It contains a great amount of polyrhythms (multiple simultaneous tempo layers) and polymeters (multiple simultaneous meters).

It was exactly these kinds of rhythmical elements that Stravinsky used in *Le Sacre du Printemps;* rhythms that, despite their complexity, somehow felt natural and appealing (and tremendously provocative at the time). How come that complexity in this case feels so natural and organic? Could it have something to do with the social aspects that somehow are integrated into the DNA of African traditional music? Or is it due to pure musical reasons detached from any social aspects? If this is the case, why does the human ear find some types of complex rhythms highly organic and others awfully constructed and lifeless?

These questions are not easy to answer. However, we might be wiser on the social aspects of rhythm by tracing down rhythm to its origins. African music culture is one of the oldest that we know of, and luckily the origins are still very present in traditional African music. Before we go deeper into the subject, I would first like to clear out some exciting technicalities.

4.7   Tempo

Rhythm alone doesn't belong to anyone. It is an independent phenomenon. No instrument has a monopoly on rhythm. Nor does man have the monopoly. Rhythm exists in cosmos, but also inside every living creature on planet Earth. It can be generated by computers as well as woodpeckers. As a human, however, it is rather difficult to separate rhythm from being human. What may be a rhythm to us may be something completely different to extraterrestrial beings and vice versa. As earlier described, rhythm derives from pulsation. So whether we want to play rhythms, listen or dance to them, we would first of all need some kind of relatable tempo as a base. We need a tempo that we can *feel*. Not just hear, but feel. And as humans, we can only perceive tempo within certain frames.

Fastest. If a tempo is extremely fast, it will be hard to distinguish between the events that we hear. Technically, we can register rhythm up to about 600 bpm (beats per minute). If the tempo gets faster than this, it will just sound like a blur of notes rather than any sort of discernible rhythm. And if the tempo goes even higher – up to about 1200 bpm – we will suddenly hear a continuous note with clear pitch and not rhythm at all. A metronome mark at 1200 bpm can be translated into 20 hertz (or 20 cycles per second). Pitch is measured in hertz, and 20 hertz is about the lowest that a human can hear. This means that a steady rhythm and pitch are exactly the same thing – it's just the human ear that perceives them differently when the speed changes.

Slowest. According to musicologist Paul Fraisse, the slowest tempo that humans can feel is about 33 bpm. If a tempo gets much slower than that, we can *hear* the beats, but it gets extremely hard to *connect* them. Again, it might be due to our human body. Try to set a metronome at tempo 25 and walk to it. Simply synchronize your steps to the clicks. Well, you'll experience that it's not so simple at all. Our slowest (natural) walking speed tells a lot about how slowly we can connect two (or more) sonic events without losing the feeling of rhythm. Walking very slowly is just extremely difficult. Not so strange that the Italian language uses the word *andante* (walking) as an indication of a musical tempo. At an extremely slow tempo like, e.g., 25 bpm, your brain might remember the previous beat, but your body would simply not be able to feel the relation of beats as coherent. Raising the tempo, the feeling of pulse becomes present though. At about 33 bpm or more, each beat will spontaneously leave a *time point* in the mind of the listener, and the feeling of pulse will also be projected *forward* in time, creating a row of anticipated time points. This feeling of past beats combined with the expectation of future beats will guide the synchronization of the body movements of anyone who may play or dance.

4.8   Chasing the Roots

We've now reached an interesting point: rhythms and body movements. Physical movement is inseparable from playing as well as (of course) dancing. No matter whether we hit a drum or dance to a groove, physical movement is a necessary part of the action. In traditional African music, the dancers often tie bells around their forearms and angels so they will sound in time with their steps and movements (often singing at the same time). In this manner, the barrier between musicians and dancers is drastically minimized. Music becomes physical. And since anyone may participate, one way or another, it also becomes social. According to the ethnomusicologist Simha Arom, there isn't even such a thing as "professional" musicians in (Central) Africa. Music is created by everyone, which means there is no clear distinction between musicians and non-musicians (of course, there are master musicians etc., but that's another matter). Due to the lack of barriers between performers and audiences, music becomes a thing that connects the community on a daily basis. Music, rhythm, and dance are not just for pleasure. It's a *need*. If one imagines the concept of rhythm as an indispensable part of everyday life – as something necessary to maintain physical and social balance, we are most likely approaching something that could be the essence of the origins of rhythm on this planet; a universal force that is projected into human life. And most importantly, that binds us together socially and physically.

Although the origins of rhythm may be of social and physical character, this is not necessarily the ultimate end goal of rhythmic purpose. The human use of rhythm has evolved. Rhythm is now all over the planet and is used in a huge variety of ways with the most incredible outcomes. No doubt that the unifying function still is an important part of rhythm. But there are so many other things that rhythm is capable of.

4.9   Methods and Training

In Africa, musical schools or education systems that guide the way into traditional music are very few. Instead, music is learned the same way as children learn how to speak, and traditions are passed on from one generation to another. Wrapped in carrying clothes, the baby is fixed onto its mother's back and thus participates in daily activities, including ceremonies and dances where the baby is "being danced" by its mother long before it can even walk. In this way, the child absorbs in the most natural manner possible the fundamentals of the music of its community.

From a music pedagogical point of view, this physical and natural way of learning music has incredibly good results, as playing becomes just as natural as walking or speaking. But there are other educational ways to approach rhythm than this. Let's take a look at Indian classical music. Here is the way of understanding and learning rhythms completely different. And with results too amazing to be described in words. In Indian classical music, the system of rhythm is extremely calculated. And the musical education system is remarkably organized. It is common for students to visit their gurus daily to learn music, but it typically takes several years before a student is "skilled" enough to perform in public. This is a complete contradiction to the African way in which the child starts to imitate adults and older kids at about walking age and then joins the musical community shortly after. In India, rhythm training is an obligatory part of music lessons, no matter what instrument the student is playing. The rhythm training is based on a system that is highly methodical. Let me shortly describe how a polyrhythm can be generated using the South Indian system.

In India, rhythm is understood by subdividing the pulse into various units, like slicing up a cake (I'm using western notation and not the Indian notation system).

 etc.

The notes in the top line indicate the pulse, while the lower line shows subdivisions from 2 to 8. This kind of measured rhythmical system is quite related to the western system, which tends to measure everything mathematically.

Now let's go a little further. If we pick one of the division types above, e.g. 5 units per beat (the quintuplet), but mark every (e.g.) 4th unit, the outcome would look like this:	Or simply notated in this way, this emphasizes the consistent phrasing of 4-groups inside quintuplets (the top layer shows the pulse).
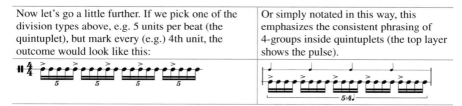	

If the unaccented beats were removed, the polyrhythm 5:4 would stand alone. Let's take the rhythm a little further out by replacing every 4-group with a 7-group (a septuplet):

The total outcome of the cycle above is a 35:4 polyrhythm (7 × 5 over 4). Already a fairly complex tempo constellation. But still, it feels a little stiff musically, so let's loosen it up by randomly removing some of the units in the septuplet layer.

Suddenly the rhythm comes alive. With these omissions, the rhythm becomes much more fun to listen to with its playfulness and syncopations. However, the most notable aspect of this rhythm is the fact that the quite sophisticated texture is played *inside* a polyrhythm. Playing this rhythm, the performer is basically dealing with two simultaneous layers of tempo – something that is perfectly possible, but which usually requires years of specific training.

4.10   The Western Explosion

Even though the bar above would be very difficult for almost any western musician, the exact same bar would be played instantly if it was presented to any South Indian musician. This kind of rhythmical texture has been a part of Indian music since approximately 500 BC. Compared with the western music tradition, rhythms with a corresponding level of complexity were not seen until the middle of the twentieth century. But since then, the way of using rhythms in western contemporary music has exploded. Composers like Conlon Nancarrow, Iannis Xenakis, and Brian Ferneyhough have written music with rhythmic textures out of this world. Conlon Nancarrow wrote music for auto-playing pianos because his rhythms were too difficult for anyone to play. Iannis Xenakis used his mathematical background as an architect to generate massively intense rhythm. Brian Ferneyhough uses tools similar to Indian systems in order to create rhythms inside rhythms inside rhythm. Like rhythmic Babushka dolls. Just check out the complexity level of this excerpt from his *Bone Alphabet* for solo percussion:

Edition Peters No. 7389

These three composers, and thousands of others, use the art of rhythm to create music of any conceivable kind. In some works, rhythm is used as a tool to create chaos. In other works, rhythm may produce order and stability. Sometimes rhythm makes the music appear dangerous. Other times pleasant. Sometimes groovy. Other times shady. Sometimes free and jazzy. Other times strict and constant. Sometimes supernatural. Other times natural. Rhythm works in many and mysterious ways.

When it comes to complexity in rhythm, the outcome wouldn't necessarily sound strange and complicated. In my opinion, some extreme rhythms sound very organic and natural. Earlier I described how tempo and tone (pitch) are the same – just experienced differently when speed changes.

Below it's clear how musical intervals can be described by simple ratios. Since the bottom note vibrates at 440 hertz, it relates to the top note by a ratio of 3:2. Simple math.	The ratio 3:2 can, as we know, also be used to describe a polyrhythm. If speeded up to tempo 13,200 bpm, this rhythm would sound exactly like the interval to the left. Simple physics.
$$\frac{660hz}{440hz} = \frac{3}{2}$$	

The numbers 2 and 3 are the most fundamental building blocks of music. Let's try to add an extra layer to the interval above. By adding a *contrasting* number to the constellation of 2 and 3, the interval would turn into a harmony. If 4 is added, it wouldn't do any good for the interval since 4 just is a double of 2. But 5 would be miraculous. It's the first and finest exception to 2 and 3.

The vibration speeds 1100 Hz, 660 Hz, and 440 Hz can be described by the ratio 5:3:2. If played simultaneously, a major triad would occur. The major triad is the simplest existing harmony because it's the product of the simplest (contrasting) ratios.	Translated into rhythm, the same ratios would look like this.
$1100 hz : 660 hz : 440 hz = 5 : 3 : 2 =$	

It's fascinating that a polyrhythm is a harmony in slow motion. And even more fascinating is that the harmony and rhythm actually *feel* the same. Strong, open, natural, and noble are some of the words that I would use to describe the harmonic and rhythmic product of 2:3:5. Despite its organic flow, the 2:3:5 polyrhythm is still considered a fairly advanced rhythm by most western musicians.

Returning to the question of why Stravinsky's rhythms in *Le Sacre du Printemps* felt outlandish yet so natural, the natural and organic aspect may have to do with the fact that the rhythmical structures are built upon simple ratios like the rhythms above (although they may not look simple). Ratios are found all around us in nature, as well as in the art of painting and architecture. But why did the rhythms provoke as much as they did? Well, simply because they had never been heard before in western music when the work was premiered in Paris in May 1913. However, the very same rhythmic ratios had been part of music from other cultures for millennia.

No matter how extreme a rhythm may appear, it can always be traced down to the simplicity of pulsation. And that might be the reason why rhythm in its nature has a unifying effect on humans. No matter which country or culture one may belong to – or what religious, non-religious, or political belief one may have, the phenomena of pulsation is simply relatable to anyone on the planet, simply because rhythm exists all around us. It's everywhere. It's in cosmos – and it's in the heart of each and every one of us.

Suggested Reading

Arom S. African polyphony and polyrhythm. Cambridge: Cambridge University Press; 1991.
Bernstein L. The unanswered question. 1973; https://www.youtube.com/playlist?list=PLFjonLo8 gYHIXC35K4Ujrbu6XHchNDCv9
Bolton TL. The American journal of psychology. Champaign, IL: University of Illinois Press; 1894.
Cohn R. How music theorists model time. J Music Theory. 2021;
Cowell H. New musical resources. Cambridge: Cambridge University Press; 1930.

Fraisse P. The psychology of time. Manhattan, NY: Harper & Row; 1963.

Handel S. Using polyrhythms to study rhythm. Berkeley, CA: University of California Press; 1984.

Mathieu WA. Harmonic experience. Rochester, VT: Inner Traditions International; 1997.

Neely A. https://www.youtube.com/c/AdamNeely

Reina R. Applying karnatic rhythmical techniques to western music. Milton Park: Routledge; 2015.

Tecumseh Fitch W. The biology and evolution of rhythm: unravelling a paradox. Laurinburg, NC: University of St Andrews; 2011.

A Lullaby to the Brain: The Use of Music as a Sleep Aid

5

Kira Vibe Jespersen

5.1 Lullabies and the Use of Music for Sleep

The use of music for soothing babies and singing them to sleep is one of the most prevalent uses of music in all societies across the globe [1]. This common use of music to facilitate sleep in children is reflected in the music genre of lullabies. Despite cultural differences, lullabies seem to share common features such as a slow tempo and a simple rhythmical and melodic structure [2, 3]. As such, even culturally unfamiliar lullabies can be consistently identified by naïve listeners and correctly distinguished from, e.g., healing songs and love songs [2, 4]. Studies also show that infants respond with relaxation even to unfamiliar lullabies [5], suggesting that this reaction is not a function of their musical experiences but rather a predisposed response to the universal features characterizing lullabies.

With the enormously increased accessibility to recorded music within the last decades, music is now also commonly used by adults when they have trouble sleeping. In a large Canadian survey study, they found that 25% of the population reported listening to music at bedtime when experiencing sleep difficulties [6], and this number increased to 43% among individuals suffering from insomnia disorder. Similarly, a survey of sleep habits among university students revealed that listening to music was reported as a common strategy to improve sleep by 55% of the responders [7]. The common use of music as a tool to improve sleep is also reflected in the large number of sleep playlists found on global streaming services such as Spotify [8].

K. V. Jespersen (✉)
Department of Clinical Medicine, Center for Music in the Brain, Aarhus University, Aarhus, Denmark
e-mail: kira@clin.au.dk

© The Author(s), under exclusive license to Springer Nature Switzerland AG 2022
B. Colombo (ed.), *The Musical Neurons*, Neurocultural Health and Wellbeing, https://doi.org/10.1007/978-3-031-08132-3_5

5.2 Music to Alleviate Sleep Problems

The existence of the music genre lullabies and the common use of music as a self-help strategy to alleviate sleep problems suggest that listening to music at bedtime could potentially be a beneficial intervention for people with sleep problems. Sleep problems are very common with around 1/3 of the general population experiencing insomnia symptoms such as difficulties falling asleep or maintaining sleep [9]. Insomnia is more common in women than men, and the prevalence seems to increase with age [10]. As such, music listening could be a potentially relevant intervention as it is safe, easy to use, and low-cost compared to other treatment options.

A number of clinical trials have been conducted, and systematic reviews have summarized the effect of music as a sleep aid. A Cochrane review from 2015 investigated the effect of music on sleep quality in adults with sleep difficulties [11]. Through a systematic literature search, six studies were identified that assessed the effect of listening to music on sleep in adults with a complaint of poor sleep. Five of these studies could be included in a meta-analysis, and the results showed a consistent beneficial effect of music listening on sleep quality [11]. Recently, more studies have investigated the effect of music as a sleep aid in various populations with sleep problems.

5.2.1 Age-Related Sleep Problems

A number of clinical trials have examined the effect of music on improving sleep in elderly persons with sleep problems, and three systematic reviews have summarized the effect using slightly different criteria. One systematic review included only randomized controlled trials and identified five studies on adults older than 60 with sleep difficulties [12]. When synthesizing the results in a meta-analysis, they found improvement in sleep quality with the music intervention compared to the control group. The results suggested that listening to sedative music was more effective than rhythm-centered music, and that an intervention period longer than 4 weeks was associated with a greater effect than shorter intervention periods. These results were supported by another systematic review finding similar results in their meta-analysis [13]. However, a third review reported a less clear effect of music interventions on age-related sleep problems [14]. This review included both randomized and quasi-experimental studies, and they identified a total of 16 studies of which 11 included music listening and 5 used multi-component interventions. Of the nine randomized controlled trials investigating music listening interventions, three studies showed no effect, whereas six studies found a beneficial effect of the music on sleep.

In addition to sleep difficulties during normal aging, a couple of studies have also investigated dementia-related sleep problems. In a pilot study by Weise and colleagues, elderly persons with dementia living at a nursing home were offered individualized playlists that they listened to for 30 min every other afternoon for 4 weeks. The music intervention was successfully implemented, and it was found to facilitate improved sleep quality in the participants [15]. Another study included 40 participants listening to a sleep playlist of their preferred genre every night at bedtime for 14 days [16]. In this study, the music intervention was also well-liked, and sleep improvements were observed in half of the participants during the intervention period compared to a baseline measurement period. Total sleep time and sleep efficiency were measured with wrist actigraphy, but the results showed no changes in these sleep parameters [16]. Overall, these studies provide promising results for the use of music interventions for sleep improvements in dementia, but larger studies using high-quality methods are still needed to determine the effect.

5.2.2 Insomnia Related to Mental Disorders

Another line of research has evaluated the effect of music on sleep problems related to mental disorders. Insomnia symptoms are very common in a number of mental disorders, and studies have evaluated the impact of music on sleep in adults with depression, schizophrenia, and trauma. In a study from 2009, Deshmukh and colleagues assessed the effect of listening to music compared to hypnotic medication in adults with major depressive disorder [17]. The study was conducted in India, and the participants in the music group listened to Indian classical music, raagas, selected for sleep. After the one-month intervention period, the results showed a beneficial effect in the music group comparable to the effect seen in the control group receiving hypnotic medications [17]. These findings suggest that music may be a useful adjuvant in the management of depression-related sleep difficulties.

Insomnia symptoms are also very common in schizophrenia, and the effect of music on sleep improvement has been studied in this population [18]. In a within-subject study, 24 participants with schizophrenia listened to relaxation music composed for the study for 40 min every night at bedtime for 7 days. Sleep was measured objectively with an actigraph, which is a wrist-worn accelerometer from which you can derive the sleep–wake pattern. The results showed that participants fell asleep faster during the music intervention (shorter sleep latency), and that they slept a larger proportion of the time spent in bed (improved sleep efficiency) [18]. These findings point toward beneficial effects of relaxation music to improve sleep in schizophrenia, but until now, no further studies have been conducted.

Another group of studies have focused on sleep problems related to psychological trauma and PTSD. The first study from 2005 evaluated the effect of music listening on sleep problems experienced by abused women in shelters [19]. This study used participant-selected music together with progressive muscle relaxation for 20 min daily compared to 20 min of rest in silence. The intervention was used for five consecutive days, and the results showed a significant sleep improvement in the intervention group, but not in the control group [19]. Similarly, another study focused on sleep problems in traumatized refugees [20]. In this study, participants in the intervention group listened to researcher-chosen relaxation music for 60 min every night at bedtime for 3 weeks. The findings showed a significant reduction of sleep problems in the participants listening to music compared to the non-music control group [20]. A third study used a within-subject design to compare music listening to progressive muscle relaxation in 13 adults with PTSD and chronic insomnia [21]. The results showed improved sleep quality as reported in question-naires with the music intervention, as well as shorter sleep latency and improved sleep efficiency as measured with actigraphy [21]. Seen together, these three studies point toward a beneficial effect of music listening for trauma-related sleep prob-lems. However, it is worth noting that all three studies had small sample sizes (13–28 participants), and none of them used a randomized controlled trial design. Overall, the studies on insomnia related to mental disorders reveal the promising potential for music interventions to improve sleep, but larger studies using high-quality research designs are needed to fully determine the effect.

5.2.3 Music for Sleep Improvement in Hospitalized Patients

The effect of music on sleep improvement has also been investigated in hospitalized patients with sleep problems. Sleep is often severely disturbed during hospitaliza-tion due to factors such as acute illness, unfamiliar environment, pain, and noise [22]. A number of studies have been conducted in this field including patients in the intensive care unit [23, 24], patients undergoing cardiothoracic surgery [25–27], and patients undergoing other medical procedures [28, 29]. A systematic review has summarized the results of five randomized controlled trials including 259 patients [30]. Most of the included studies used researcher-chosen music that patients lis-tened to between 30 and 53 min, and the meta-analysis showed moderate-quality evidence for a 27% increase in sleep quality with the music intervention compared to controls [30].

In summary, there is a growing body of evidence suggesting the beneficial effects of listening to music for improving sleep in various populations. Importantly, none of the intervention studies have reported any negative side-effects of the music intervention, suggesting that listening to music is a safe intervention to use for sleep improvement [31]. However, one study has raised concerns that for some people music may trigger involuntary musical imagery (earworms) that can be sleep disruptive [32]. This study used participants prone to involuntary musical imagery as well as music stimuli known to trigger earworms, so it remains unclear if this is an issue in populations with sleep problems when using music chosen for sleep.

5.3   How Can Music Facilitate Sleep?

As seen in the previous section, many people use music as a tool to improve sleep, and research supports the beneficial effects of music for improving sleep quality. But how can music affect our sleep? In the last decades, research in music neurosci-ence and music psychology has expanded greatly, and some of the results may help us understand the mechanisms underlying the impact of music on sleep. When considering the mechanisms underlying the effect of music as sleep aid, it is important to take the nature of the sleep problems into account. Poor sleep can be related to a number of factors, including hyperarousal, negative emotional states, or external factors such as noise. We know that music can engage various human functions [33], and music as an intervention to improve sleep may therefore work through several different mechanisms. Therefore, we will look at the relevant research focusing on brain mechanisms, bodily responses, psychological mechanisms, and environmental factors.

5.3.1 The Brain Mechanisms of Sleep Music

Sleep is essential for brain health, and during sleep the neural activity changes and enters a cycle of REM (rapid-eye movement) sleep and non-REM sleep that can be divided into three stages of lighter and deeper sleep (N1, N2, and N3) [34]. N1 corresponds to the light sleep when you fall asleep, N2 is more consolidated sleep, and N3 is slow-wave sleep, also termed deep sleep. A few experimental and clinical studies have investigated the impact of music on these different sleep stages. One clinical study measured the effect of a four-day music intervention on adults with sleep complaints in a sleep laboratory using polysomnography, which is the gold standard of sleep assessment including electroencephalography (EEG), electrooculography (EOG), and electromyography (EMG). They found shortened N2 sleep and prolonged REM sleep in the music group compared to the control group. However, the size of the effect was small [35]. Another study used one-channel EEG to measure sleep in the homes of the participants who were adults with poor sleep [36]. Compared to music video and a no-intervention control group, they found no change in objective sleep measures after the four-day music listening intervention. Similarly, Jespersen and colleagues found no change in sleep architecture in a clinical study focusing on adults with insomnia disorder [37]. In this study, they measured sleep with ambulant polysomnography before and after a three-week intervention period by comparing music listening to audiobooks and a waitlist control group. The finding showed no difference between the groups in objective sleep parameters [37].

In addition to these clinical studies, one experimental study used a within-subject design to evaluate the immediate effect of listening to relaxation music compared to listening to spoken text in healthy participants during an afternoon nap [38]. The results showed that N1 sleep was reduced with 8 min in the music condition. When dividing participants into low and high scores on a scale of hypnotic susceptibility, they found that participants with low susceptibility showed an 46% increase in slow-wave sleep (N3) with the music condition compared to the text condition. However, this effect was not found in participants with high susceptibility scores [38]. Overall, these studies show mixed results of the impact of music on objective sleep measures, such as the amount of time spent in the different sleep stages. Very few studies have investigated these underlying aspects of sleep, and it remains unclear if objective changes in the neural activity patterns during sleep underlie the experienced improvement in sleep quality. Hopefully, future studies will shed more light on the degree to which music can affect the various sleep stages.

5.3.2   Bodily Responses to Sleep Music

Another important aspect of sleep is the physiological changes associated with the transition into sleep. When falling asleep, we relax throughout the body, our breath gets slower, the heart rate is reduced, and our blood pressure goes down [34]. We know from experimental research that music can affect these bodily responses, and this may also be one of the mechanisms underlying the impact of music on sleep.

To investigate the stress-reducing potential of music, de Witte and colleagues reviewed randomized controlled trials examining the effect of music on physiological measures such as heart rate and blood pressure [39]. They found 61 relevant studies including 3188 participants. These studies included the use of music as an intervention compared to usual care or other interventions in a number of clinical and nonclinical settings. The results of the meta-analysis showed a small-to-medium effect (Cohen's $d = 0.380$) of the music on physiological outcomes [39]. Moderator analyses showed that the effect was larger for heart rate (Cohen's $d = 0.456$) compared to blood pressure (Cohen's $d = 0.343$) and stress hormone levels (Cohen's $d = 0.349$). These results suggest that music can reduce physiological arousal, and this may be an important mechanism for facilitating sleep.

As seen in the lullaby research, the bodily responses are closely linked to the characteristics of the intervention music. For example, breathing rate, heart rate, and blood pressure have been shown to increase with fast or complex music in contrast to slow, predictable music [40]. Similarly, an experimental study showed that music dynamics and rhythmic intensity were tracked consistently in physiological measures of cardiovascular and respiratory activity [41]. Therefore, the music characteristics are important to consider when choosing music for sleep improvement.

5.3.3   Psychological Mechanisms Underlying the Impact of Music on Sleep

In the previous sections, we have seen how the impact of music on sleep may relate to neural mechanisms and bodily responses to music. However, we also know that psychological aspects are essential when it comes to both music and sleep. Sleep difficulties are often associated with difficult emotional states such as depression and anxiety [10]. Furthermore, many people with insomnia experience heightened arousal and racing thoughts at bedtime. Therefore, research on the emotional impact of music may be relevant to understanding how music can affect our sleep.

A strong line of research documents the potential of music for affecting human emotions, and many people rate music among the top 10 pleasures in their lives

[42]. Music is commonly used for mood regulation through mechanisms such as mood improvement, distraction, and relaxation [43]. Mood improvement may be the result of pleasurable responses to well-liked music, and neuroscience studies have shown that music can engage human reward networks and thereby elicit pleasurable responses and mood improvement [44, 45]. Improved mood may be one important factor in the impact of music on sleep, especially for people experiencing insomnia in relation to mood disorders or anxiety.

In addition to the pleasurable responses to music, mood regulation can also be achieved through distraction. Studies show that distraction is a common strategy in the use of music for regulating emotions and mood [43]. Music can catch our attention and divert us from troublesome feelings, thoughts, or worries. Thereby, this mechanism may be particularly relevant to people who experience bedtime rumination or racing thoughts. Music can also distract us from pain or other unpleasant physical sensations and thus ease the experience of discomfort [46]. This effect can be relevant in relation to insomnia related to medical conditions or pain syndromes.

Finally, music mood regulation can also happen through relaxation. People at all ages report the use of music for recovery, calming down, and taking a break [43]. As seen above, the effect of music is reflected in the bodily responses to soothing music, and this effect is also seen in the emotional responses. A meta-analysis on the impact of music on stress-related outcomes identified 79 studies with a total number of 6800 participants reporting the effect of music on psychological outcomes such as state anxiety, nervousness, restlessness, and feelings of worry [39]. The results showed a medium-to-large effect of the music interventions on these outcomes. Since anxiety, worries, and restlessness are common in insomnia, the alleviation of these feelings may also be one of the mechanisms whereby music can facilitate sleep.

5.3.4 Environmental Factors in the Impact of Sleep Music

We have seen that a number of neurophysiological and psychological mechanisms may underlie the effect of music on sleep. In addition to these, external factors related to the environment may also be relevant. In some cases, poor sleep may relate to a noisy environment. This is often true for hospitalized patients where the unfamiliar environment with sounds and 24-h activity can be severely disturbing for sleep, even though sleep is much needed for healing and recovery [30]. Other sources of noise may be traffic noise if you live in a city environment, or noisy neighbors. Here, music may facilitate sleep by masking the external noise. At the same time, some people can be uncomfortable with the silence of the night and therefore report using music to create a comfortable auditory environment and mask the silence [47].

In summary, music may facilitate sleep through a number of different neurological, physiological, psychological, and environmental pathways, even though the exact relevance of these mechanisms for sleep remains to be settled. Still, there seems to be a close link between the mechanisms and the characteristics of the music used as a sleep aid.

5.4   The Characteristics of Sleep Music

In the previous sections, we have looked at the effects of music as a sleep aid and the potential underlying mechanisms of this effect. Another question that is commonly raised is what kind of music to choose for sleep.

A couple of studies have investigated the characteristics of music used as a sleep aid through surveys or using big data from international streaming services. A UK-based survey including 651 responders found that the music people used to sleep was very diverse, including various genres and artists [47]. Therefore, the authors suggest that the choice of sleep music seemed to be driven largely by individual preferences. However, this study only collected data on the artists and genres, and the researchers did not analyze the music characteristics or audio features. Another survey study from Australia also found a large variation in the music used to improve sleep [48]. In this study, they analyzed the musical features and found that despite the variation, music used as a sleep aid also shared some common characteristics such as a medium tempo, smooth sounds (legato articulation), and low-to-medium rhythmic activity. Still, this study was based on the responses of 161 students, and therefore the results have very limited generalizability.

A recent study by Scarratt and colleagues accounted for some of these limitations by collecting big data from the international streaming service Spotify to identify both the general features of sleep music and potential subgroup characteristics [8]. They identified 225,927 tracks used for sleep, and the results showed that compared to music in general, sleep music was generally slower, less energetic, more instrumental, and more often played on acoustic instruments. Still, they also found a large amount of variation in the type of music used as a sleep aid, and interestingly, the sleep music clustered into six distinct subgroups. The largest cluster was termed "Ambient music" and consisted of slow soothing tracks with low-to-medium rhythmic saliency similar to the expected features of sleep music. However, three subgroups showed very different characteristics and consisted mainly of up-tempo pop music as well as indie, lofi, and R&B music. These three clusters shared characteristics such as the presence of lyrics as well as high energy and loudness [8]. In summary, this large global dataset points toward some general characteristics in the music people choose to listen to as a sleep aid, but also substantial variation. Thus, the results suggest a balance between universal music characteristics that may facilitate sleep and the importance of individual preferences.

Based on some of the above-mentioned mechanisms, one could argue that sleep music should be sedative to facilitate a relaxation response, i.e., with a slow tempo and low rhythmic activity [49]. This is in line with some of the characteristics found in the studies above. On the other hand, some of the emotional mechanisms may favor music that is familiar and well-liked by the user. These individual differences may be the factor that is reflected in the large variation found in sleep music. One limitation to these studies on people's individual choice of sleep music is that we do not know if the music used really improved the sleep of the users. One study has compared the effect of researcher-chosen music to self-selected music for sleep and found no difference in the effect [50]. Similarly, a meta-analysis found no difference between the effect in studies using researcher-chosen music and studies giving the participants a choice among pre-defined play-lists [11]. Thus, individual music preferences seem to play a role when people choose music to help them sleep, but the specific impact on the effect on sleep problems remains to be disentangled.

5.5   Conclusion

In this chapter, we have seen that music is commonly used as a sleep aid when people have trouble sleeping. The close link between music and sleep is reflected in the music genre lullabies, which seem to be one of the most common uses of music for a specific behavioral purpose across the globe. In adults, music can indeed facilitate sleep in various populations with sleep difficulties, even though the evidence is strongest for elderly people and sleep problems related to hospitalization. Future research should disentangle the underlying mechanisms, including the relative influence of specific music features and individual music preferences. Poor sleep can impair both physical and mental health, and music could be a safe, viable, and low-cost tool to improve sleep for the millions of people suffering from insomnia.

References

1. Mehr SA, Singh M, Knox D, Ketter DM, Pickens-Jones D, Atwood S, et al. Universality and diversity in human song. Science. 2019;366(6468):944.
2. Mehr SA, Singh M, York H, Glowacki L, Krasnow MM. Form and function in human song. Curr Biol. 2018;28(3):356–68.e5.
3. Trehub SE, Trainor L. Singing to infants: lullabies and play songs. Adv Infancy Res. 1998;12:43–78.
4. Trehub SE, Unyk AM, Trainor LJ. Adults identify infant-directed music across cultures. Infant Behav Dev. 1993;16(2):193–211.
5. Bainbridge CM, Bertolo M, Youngers J, Atwood S, Yurdum L, Simson J, et al. Infants relax in response to unfamiliar foreign lullabies. Nat Hum Behav. 2021;5(2):256–64.
6. Morin CM, LeBlanc M, Daley M, Gregoire JP, Mérette C. Epidemiology of insomnia: prevalence, self-help treatments, consultations, and determinants of help-seeking behaviors. Sleep Med. 2006;7(2):123–30.
7. Brown CA, Qin P, Esmail S. "Sleep? Maybe later…" a cross-campus survey of university students and sleep practices. Edu Sci. 2017;7(3):66.
8. Scarratt RJ, Heggli OA, Vuust P, Jespersen KV. The music people use to sleep: universal and subgroup characteristics. PsyArXiv. 2021.
9. Ohayon MM. Epidemiology of insomnia: what we know and what we still need to learn. Sleep Med Rev. 2002;6(2):97–111.
10. Morin CM, Benca R. Chronic insomnia. Lancet. 2012;379(9821):1129–41.

11. Jespersen KV, Koenig J, Jennum P, Vuust P. Music for insomnia in adults. Cochrane Database Syst Rev. 2015;(8):CD010459-CD.
12. Chen CT, Tung HH, Fang CJ, Wang JL, Ko NY, Chang YJ, et al. Effect of music therapy on improving sleep quality in older adults: a systematic review and meta-analysis. J Am Geriatr Soc. 2021;69(7):1925–32.
13. Wang C, Li G, Zheng L, Meng X, Meng Q, Wang S, et al. Effects of music intervention on sleep quality of older adults: a systematic review and meta-analysis. Complement Ther Med. 2021;59:102719.
14. Petrovsky DV, Ramesh P, McPhillips MV, Hodgson NA. Effects of music interventions on sleep in older adults: a systematic review. Geriatr Nurs. 2021;42(4):869–79.
15. Weise L, Töpfer NF, Deux J, Wilz G. Feasibility and effects of individualized recorded music for people with dementia: a pilot RCT study. Nord J Music Ther. 2020;29(1):39–56.
16. Jespersen MJ, Vuust P. Bedtime music for sleep problems in older adults with dementia: a feasibility study. Music Med. 2020;12(4):222–30.
17. Deshmukh AD, Sarvaiya AA, Seethalakshmi R, Nayak AS. Effect of Indian classical music on quality of sleep in depressed patients: a randomized controlled trial. Nord J Music Ther. 2009;18(1):70–8.
18. Bloch B, Reshef A, Vadas L, Haliba Y, Ziv N, Kremer I, et al. The effects of music relaxation on sleep quality and emotional measures in people living with schizophrenia. J Music Ther. 2010;47(1):27–52.
19. Hernández-Ruiz E. Effect of music therapy on the anxiety levels and sleep patterns of abused women in shelters. J Music Ther. 2005;42(2):140–58.
20. Jespersen KV, Vuust P. The effect of relaxation music listening on sleep quality in traumatized refugees: a pilot study. J Music Ther. 2012;49(2):205–29.
21. Blanaru M, Bloch B, Vadas L, Arnon Z, Ziv N, Kremer I, et al. The effects of music relaxation and muscle relaxation techniques on sleep quality and emotional measures among individuals with posttraumatic stress disorder. Ment Illn. 2012;4(2):59–65.
22. Altman MT, Knauert MP, Pisani MA. Sleep disturbance after hospitalization and critical illness: a systematic review. Ann Am Thorac Soc. 2017;14(9):1457–68.
23. Hansen IP, Langhorn L, Dreyer P. Effects of music during daytime rest in the intensive care unit. Nurs Crit Care. 2018;23(4):207–13.
24. Su CP, Lai HL, Chang ET, Yiin LM, Perng SJ, Chen PW. A randomized controlled trial of the effects of listening to non-commercial music on quality of nocturnal sleep and relaxation indices in patients in medical intensive care unit. J Adv Nurs. 2013;69(6):1377–89.
25. Hu R-F, Jiang X-Y, Hegadoren KM, Zhang Y-H. Effects of earplugs and eye masks combined with relaxing music on sleep, melatonin and cortisol levels in ICU patients: a randomized controlled trial. Crit Care. 2015;19(1):115.
26. Ryu M-J, Park JS, Park H. Effect of sleep-inducing music on sleep in persons with percutaneous transluminal coronary angiography in the cardiac care unit. J Clin Nurs. 2012;21(5–6):728–35.
27. Zimmerman L, Nieveen J, Barnason S, Schmaderer M. The effects of music interventions on postoperative pain and sleep in coronary artery bypass graft (CABG) patients. Sch Inq Nurs Pract. 1996;10(2):153–70; discussion 71–74.
28. Kim J, Choi D, Yeo MS, Yoo GE, Kim SJ, Na S. Effects of patient-directed interactive music therapy on sleep quality in postoperative elderly patients: a randomized-controlled trial. Nat Sci Sleep. 2020;12:791–800.
29. Renzi C, Peticca L, Pescatori M. The use of relaxation techniques in the perioperative management of proctological patients: preliminary results. Int J Color Dis. 2000;15(5):313–6.
30. Kakar E, Venema E, Jeekel J, Klimek M, van der Jagt M. Music intervention for sleep quality in critically ill and surgical patients: a meta-analysis. BMJ Open. 2021;11(5):e042510.
31. Koenig J, Jarczok MN, Warth M, Harmat L, Hesse N, Jespersen KV, et al. Music listening has no positive or negative effects on sleep quality of normal sleepers: results of a randomized controlled trial. Nord J Music Ther. 2013;22(3):233–42.
32. Scullin MK, Gao C, Fillmore P. Bedtime music, involuntary musical imagery, and sleep. Psychol Sci. 2021;32(7):985–97. https://doi.org/10.1177/0956797621989724.

33. Särkämö T, Tervaniemi M, Huotilainen M. Music perception and cognition: development, neural basis, and rehabilitative use of music. Wiley Interdiscip Rev Cogn Sci. 2013;4(4):441–51.
34. Kryger MH, Roth T, Dement WC. Principles and practice of sleep medicine. Berlin: Elsevier; 2017.
35. Chang ET, Lai HL, Chen PW, Hsieh YM, Lee LH. The effects of music on the sleep quality of adults with chronic insomnia using evidence from polysomnographic and self-reported analysis: a randomized control trial. Int J Nurs Stud. 2012;49(8):921–30.
36. Huang C-Y, Chang E-T, Hsieh Y-M, Lai H-L. Effects of music and music video interventions on sleep quality: a randomized controlled trial in adults with sleep disturbances. Complement Ther Med. 2017;34:116–22.
37. Jespersen KV, Otto M, Kringelbach M, Van Someren E, Vuust P. A randomized controlled trial of bedtime music for insomnia disorder. J Sleep Res. 2019;28(4):e12817.
38. Cordi MJ, Ackermann S, Rasch B. Effects of relaxing music on healthy sleep. Sci Rep. 2019;9(1):1–9.
39. de Witte M, Spruit A, van Hooren S, Moonen X, Stams G-J. Effects of music interventions on stress-related outcomes: a systematic review and two meta-analyses. Health Psychol Rev. 2020;14(2):294–324.
40. Bernardi L, Porta C, Sleight P. Cardiovascular, cerebrovascular, and respiratory changes induced by different types of music in musicians and non-musicians: the importance of silence. Heart. 2006;92(4):445–52.
41. Bernardi L, Porta C, Casucci G, Balsamo R, Bernardi NF, Fogari R, et al. Dynamic interactions between musical, cardiovascular, and cerebral rhythms in humans. Circulation. 2009;119(25):3171–80.
42. Rentfrow PJ, Gosling SD. The do re mi's of everyday life: the structure and personality correlates of music preferences. J Pers Soc Psychol. 2003;84(6):1236.
43. Saarikallio S. Music as emotional self-regulation throughout adulthood. Psychol Music. 2011;39(3):307–27.
44. Koelsch S. Brain correlates of music-evoked emotions. Nat Rev Neurosci. 2014;15(3):170–80.
45. Blood AJ, Zatorre RJ. Intensely pleasurable responses to music correlate with activity in brain regions implicated in reward and emotion. Proc Natl Acad Sci. 2001;98(20):11818–23.
46. Lunde SJ, Vuust P, Garza-Villarreal EA, Vase L. Music-induced analgesia: how does music relieve pain? Pain. 2019;160(5):989–93.
47. Trahan T, Durrant SJ, Müllensiefen D, Williamson VJ. The music that helps people sleep and the reasons they believe it works: a mixed methods analysis of online survey reports. PLoS One. 2018;13(11):e0206531.
48. Dickson GT, Schubert E. Musical features that aid sleep. Music Sci. 2020; https://doi.org/10.1177/1029864920972161.
49. Grocke D, Wigram T. Receptive methods in music therapy: techniques and clinical applications for music therapy clinicians, educators and students. London: Jessica Kingsley Publishers; 2006.
50. Yamasato A, Kondo M, Hoshino S, Kikuchi J, Ikeuchi M, Yamazaki K, et al. How prescribed music and preferred music influence sleep quality in university students. Tokai J Exp Clin Med. 2020;45(4):207–13.

Music in Dementia: From Impairment in Musical Recognition to Musical Interventions

6

Federica Agosta, Maria Antonietta Magno, Elisa Canu, and Massimo Filippi

6.1 Introduction

Dementia, defined as a significant decline in cognitive functions interfering with independence in daily activities [1], is one of the most common causes of disability and dependency in the elderly. According to the World Alzheimer Report published in 2021 [2], over 55 million people worldwide currently suffer from dementia. Because of the ever-increasing life expectancy in most countries, this number is expected to double every 20 years, reaching around 139 million by 2050 (World Health Organization 2020).

Accounting for roughly 60–80% of all cases [3], Alzheimer's disease (AD) is the most common type of dementia in older subjects, causing a progressive decline in

F. Agosta
Neuroimaging Research Unit, Division of Neuroscience,
IRCCS San Raffaele Scientific Institute, Milan, Italy

Neurology Unit, IRCCS San Raffaele Scientific Institute, Milan, Italy

Vita-Salute San Raffaele University, Milan, Italy

M. A. Magno · E. Canu
Neuroimaging Research Unit, Division of Neuroscience,
IRCCS San Raffaele Scientific Institute, Milan, Italy

M. Filippi (✉)
Neuroimaging Research Unit, Division of Neuroscience,
IRCCS San Raffaele Scientific Institute, Milan, Italy

Neurology Unit, IRCCS San Raffaele Scientific Institute, Milan, Italy

Vita-Salute San Raffaele University, Milan, Italy

Neurophysiology Service, IRCCS San Raffaele Scientific Institute, Milan, Italy

Neurorehabilitation Unit, IRCCS San Raffaele Scientific Institute, Milan, Italy
e-mail: filippi.massimo@hsr.it

B. Colombo (ed.), *The Musical Neurons*, Neurocultural Health and Wellbeing,
https://doi.org/10.1007/978-3-031-08132-3_6

65

memory and eventually leading to loss of functional independence. The clinical syndromes caused by frontotemporal lobar degeneration (FTLD) are the most frequent causes of young-onset neurodegenerative dementia, with a relatively high prevalence in several countries [4]. These syndromes are divided into behavioral (bvFTD) and linguistic variants (or primary progressive aphasias, PPA), depending on their most prominent manifestations. BvFTD is characterized by abrupt behavioral and personality changes [5], while FTLD-related PPA can be further distinguished in semantic variant (svPPA), presenting progressive loss of world and semantic knowledge, and non-fluent variant (nfvPPA), defined by effortful and agrammatic speech production [6]. Although their clinical manifestations and neuropathological features are extremely heterogeneous, all these forms of dementia are due to an abnormal accumulation of proteins, both intra- and extra-cellularly, which causes failures in neuronal communication and subsequent cell death in vulnerable brain areas [7].

To date, there is no cure for dementia, and treatments can only reduce symptom progression for a limited amount of time. Only recently, Aducanumab, the first disease-modifying drug targeting neuropathology and protein build-up, has been approved by the US Food and Drug Administration for treating AD, although its efficacy is still controversial. Along with palliative pharmacological treatments, nonpharmacological approaches such as regular physical activity and cognitive stimulation have been widely employed for managing symptoms and improving patients' quality of life.

In this chapter, after a brief description of the neural substrate of music processing, we move on to discuss the most recent evidence regarding preserved and impaired musical functions in patients with dementia. We then conclude by providing an overview of the existing literature on musical interventions and their outcomes in this clinical population.

6.2   Neural Correlates of Music Processing

Processing music is an extremely complex activity, which engages a widespread circuit of cortical and subcortical brain areas.

When we listen to our favorite song, the acoustic signal coming from the inner ear is sent through the acoustic nerve to the thalamus and the brainstem. Here, anatomical connections departing to the cortex and to the limbic system give rise to the perception of musical stimuli and all their different attributes. Indeed, after receiving the auditory input from the aforementioned connections, the primary auditory cortex decodes physical information of sounds such as frequency, pitch, loudness, and individual notes. Then, the input is transferred to the surrounding auditory association areas, which elaborate more complex information like spatial location and recognition of the timbre of a voice or a musical instrument [8]. Thanks to the activity of the hippocampus, a crucial brain structure that stores our memories, we can recognize a song as familiar, or we can even start remembering an episode of our life that is particularly linked to that melody [9]. At the same time, the consonance or the dissonance of notes, the tonality, and the tempo of a melody engage our limbic system, comprising the amygdala, the orbitofrontal, and the cingulate cortices, which are responsible for the emotions we experience when listening to music [10].

Songs have an intrinsic capacity for alternating tension build-up and resolution. This creates a psychological reaction of expectancy as the melody unfolds, which is followed by a reward-like feeling when the tension is released [10]. Indeed, the dopaminergic (or reward) network is implicated in experiencing music-evoked emotions. In particular, the nucleus accumbens, a structure of this network sensitive to primary and secondary rewards [11], is considered responsible for the feeling of "chills" when the music is perceived as pleasant [10]. As the music plays, perhaps without even realizing it, we might find ourselves singing along a catchy tune or tapping the foot to the beat of the rhythm, thus activating also our sensory-motor systems, basal ganglia, and cerebellum [9],

Interestingly, elaborating perceptual, mnestic, and emotional attributes of music are partially independent abilities, as demonstrated by patients with focal brain lesions or neurodegenerative conditions who present selective impairment in one of the aforementioned processes and preservation of the others (Fig. 6.1). Several studies showed that music can elicit positive responses even in people with advanced

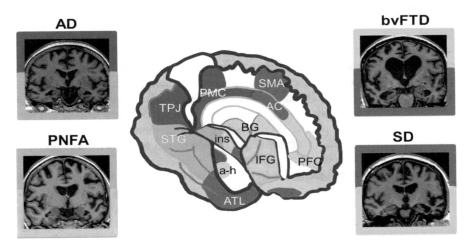

Fig. 6.1 Effects of dementia in musical cognition. Brain regions implicated in music processing are depicted and color-coded as follows: blue, tracking of musical episodic memory; green, elementary musical property (e.g., tempo) processing; cyan, scale and key processing; red, recognition of familiar musical motifs (musical semantic memory); gold, musical emotion. Coronal MRI brain sections (side panels; right hemisphere shown on the left in each section) represent patterns of atrophy typical of neurodegenerative syndromes associated with alterations in music networks. *AC* anterior cingulate; *a–h* amygdala–hippocampus; *ATL* anterior temporal lobe; *BG* basal ganglia; *IFG* inferior frontal gyrus/frontal operculum; *ins* insula; *PFC* prefrontal cortex; *PMC* posterior medial cortex (posterior cingulate, precuneus); *SMA* supplementary motor area; *STG* superior temporal gyrus; *TPJ* temporo-parietal junction. Adapted from Camilla N. Clark, Jason D. Warren, Music, memory and mechanisms in Alzheimer's disease, Brain, Volume 138, Issue 8, August 2015, Pages 2122–2125, https://doi.org/10.1093/brain/awv148; an Open Access article distributed under the terms of the Creative Commons Attribution License (http://creativecommons.org/licenses/by/4.0/), which permits unrestricted reuse, distribution, and reproduction in any medium, provided the original work is properly cited

cognitive decline [12]. Therefore, it can constitute a means of communication with patients affected by different types of dementia and a valuable tool for stimulating spared cognitive functions.

6.2.1 Dementia and Musical Perception

There is certainly a universal and unconscious propensity to impose a rhythm even when one hears a series of identical sounds at constant intervals ... We tend to hear the sound of a digital clock, for example, as "tick-tock, tick-tock" - even though it is actually "tick tick, tick tick.
   Oliver W. Sacks, Musicophilia: Tales of Music and the Brain

Perceptual integrity has rarely been the focus of research on music processing in dementia, as the majority of studies in this field investigated musical memory. Some of these studies, however, used perceptual tasks as a control condition to rule out lower-level impairments potentially hampering mnestic or affective musical processing, and they have thus contributed to our understanding of the topic.

Findings regarding pitch detection ability in AD are quite mixed. When asked to state whether two melodies that differ by one single pitch are the same or not, AD patients perform as good as healthy controls [13, 14]. These results are in contrast with other studies describing a lower performance of AD patients compared to controls in different tasks of pitch and timbre discrimination [15, 16], or in the extraction of an auditory stream from a noisy background [17]. Notably, in all of these latter studies [15–17], the perceptual auditory deficits were not "pure" but rather secondary to impairments in non-verbal working memory. This is not surprising considering the crucial role played by working memory in music processing, which by nature requires maintaining in memory an auditory information that unfolds over time.

Fewer studies investigated less frequent forms of dementia, but it appears that bvFTD and svPPA patients perform similar to both AD patients and healthy controls in a pitch discrimination task [14], while nfvPPA patients show deficits in the identification of timbre and pitch direction compared to controls [16]. Although svPPA patients were able to discriminate changes of a single pitch in unfamiliar melodies, they were impaired in detecting pitch changes in familiar songs compared to controls and AD and bvFTD patients [14]. In these patients, this finding was significantly correlated with atrophy of the right inferior and superior temporal gyrus and temporal pole, areas that are markedly vulnerable in svPPA syndrome [14].

In a more complex task requiring to state whether melodies were harmonically resolved ("finished") or not ("unfinished"), bvFTD and AD patients showed a comparably lower performance relative to controls [18]. This deficit was associated with their executive capacity or behavioral alterations [18], such as impaired ability to anticipate future events in AD [19], and impaired detection of violations in social contexts (or faux pas) in bvFTD [20]. The link between music processing and social cognition abilities was further highlighted by a recent meta-analysis, which reported compelling evidence of a neural overlap between social cognition and music perception in frontotemporal dementia [21].

Albeit mixed, these findings suggest that music perception impairments in dementia vary across the different clinical syndromes, and they might at least partially depend on the deterioration of other cognitive functions.

6.2.2 Dementia and Music-Evoked Emotions

Music imprints itself in the brain deeper than any other human experience. Music evokes emotion and emotion can bring memory. Music brings back the feeling of life when nothing else can
  Oliver Sacks

Emotions are an integral and fundamental part of musical experience. Since the beginning of times, music had the ability to bring people together, promoting sociality in all cultures, and its power in relieving tension or resonating with one's feelings, be they positive or negative, is universally recognized.

Overall, the complete [22, 23] or partial [24] sparing of emotion recognition from music excerpts in AD is reported both in group and single case studies. On the contrary, svPPA is consistently associated with impairments in identifying emotions from musical stimuli. Omar et al. [25] described the case of a 56-year-old trumpet player with svPPA who did not show any problems in musical symbol knowledge or musical memory but was unable to discriminate the correct emotions conveyed by musical excerpts, especially negative ones. Another case study shed light on the neuroanatomical correlates of such deficit, comparing two svPPA patients with either bilateral or left-predominant anterior temporal atrophy [26]. Interestingly, the emotion recognition impairment was only found in the patient with bilateral atrophy, suggesting the presence of a potential right hemispheric specialization for such tasks [26]. These findings were also replicated in group studies, which found worse emotion recognition performance in svPPA patients compared to controls [27] and AD patients [28]. Besides, svPPA patients considered melodies less pleasing than both AD patients and healthy controls [18], demonstrating a reduced music-evoked reward feeling.

Emotion recognition deficits in bvFTD are extensively described in literature. Several studies suggest that these affective impairments also occur during emotion recognition from music [18, 27, 29]. Nonetheless, a case of a musician with bvFTD who maintained good affective ability has also been described, suggesting the existence of significant inter-individual variability [30].

Overall, we can conclude that emotion recognition from music can be relatively spared in AD. On the other hand, dementias belonging to the FTLD spectrum show affective problems that are not limited to concrete visual stimuli but also embrace more abstract emotional stimuli conveyed by music. The affective dysfunctions highlighted by musical emotion recognition tasks are in keeping with the typical behavioral disturbances seen in these patients, such as apathy and lack of empathy.

6.2.3 Dementia and Musical Memory

We see a frail and elderly woman in a chair, her eyes downcast. She motions for the music to be turned up, a swelling melody from Tchaikovsky's Swan Lake, and with a little encouragement her hands begin to flutter. […]. She leans forward, wrists crossed in classic swan pose; her chin lifts as if she's commanding the stage once more, her face lost in reverie.
Windship L. 2020, November 13.
Viral video of ballerina with Alzheimer's shows vital role of music in memory. The Guardian

The scene depicted above describes Marta Gonzalez, a former ballet dancer with advanced AD who started reproducing movements learned many years before as soon as the music of the Swan Lake played. Cases like this are not new in the literature focusing on dementia and musical memory. In fact, even though AD is by definition a progressive disturbance of memory, not all memory components are equally affected. While episodic (or explicit) memories are the first ones to be compromised, procedural (or implicit) memory, the one allowing us to remember how to ride a bike forever, remains relatively preserved even in more advanced AD stages [31]. This notion is particularly evident when considering music production abilities, as reported in case studies of advanced AD patients perfectly capable of playing musical instruments [32, 33].

The integrity of explicit forms of musical memory in AD, entailing recognition of familiar music, is a more controversial issue. While some studies describe good performances in tasks of familiar and unfamiliar music recognition even in more severe AD cases [34, 35], others do not support this finding [36]. Literature is even more inconsistent when considering music recognition in non-AD dementias, as the little evidence available is mainly anecdotal and focused on expert musicians. These studies suggest that musical recognition might be preserved both in svPPA [25] and bvFTD [30].

Music has also been appreciated for its ability to evoke autobiographical memories in people with dementia. AD patients are shown to retrieve as many autobiographical memories as healthy controls in response to music [35], even more than the memories they retrieved after seeing photographs [37]. Such music-evoked autobiographical memories are not found in bvFTD patients [38], arguably due to the degeneration of medial frontal regions that are implicated in autobiographical memory. Notably, these areas are affected in bvFTD but relatively spared in AD.

It appears that dementia does not prevent patients from learning new musical information, even in the presence of severe episodic memory impairment. Cowles et al. [33] described a violinist with AD who was able to learn new songs in spite of her extremely pervasive deficit in anterograde memory. Similar results were also found in a 91-year-old non-musician woman with severe AD who successfully learned a new song [39]. Furthermore, several studies reported the emergence of musical skills in bvFTD patients who were not musically trained before disease onset [40, 41]. Such musicophilia, or an increased interest and music-seeking behavior, was found to be even more prominent in svPPA compared to bvFTD [42].

Overall, these findings indicate that memory abilities, especially implicit ones, are relatively preserved in different forms of dementia, and music can facilitate the emergence of autobiographical memories and new musical skills.

6.3   Musical Interventions in Dementia

The expression "musical interventions" broadly refers to the study and the use of music to obtain beneficial effects for individuals. When applied to clinical settings, the use of music-based protocols to accomplish therapeutic goals is more specifically known as "music therapy." Overall, this type of intervention may range from informal to more structured protocols, consisting of receptive (e.g., music listening) or active exercises (e.g., music playing, singing, dancing).

The therapeutic effects of music in dementia can be attributable to a variety of reasons. First, as discussed in the previous sections, some musical functions remain preserved in patients with dementia despite their progressive cognitive decline. Indeed, the neurodegenerative process typically affects some brain regions while others remain relatively intact for longer periods, allowing therapists to act upon the residual cognitive abilities and strengthen them. Second, other benefits may arise from the nature of music per se, given its ability to affect cognition and mood in several ways. As exemplified in a recent model [43], music is engaging in the way that both active and passive activities involve numerous functions and brain regions simultaneously, therefore promoting neural plasticity. Furthermore, the emotional connotation of music plays a huge part in the efficacy of music therapies since certain melodies can elicit positive moods and provide the key to accessing significant autobiographical memories. Indeed, music has a strong personal meaning, and it can help patients retrieve a sense of self [43]. Group sessions can also be employed to boost social support and prevent isolation.

The implementation of therapeutic protocols taking into account all these potentialities showed encouraging results on global cognition, attention, and verbal fluency [43]. For example, weekly sessions of music listening or singing over a time period of 10 weeks in patients with early dementia (diagnoses were not specified) were associated with stable or improved performances in global cognition, attention, and executive functions compared to standard care [44]. However, such positive consequences on cognition were not found in more advanced stages of dementia [45].

Although heterogeneity exists across studies [12], memory and language functions appear to benefit from musical interventions. For example, mild AD patients recalled better verbal material learned through sung lyrics after 4 weeks compared to material learned in a speaking mode [46]. Similarly, there is evidence of nfvPPA patients who are incapable of conducting a conversation but are perfectly able to sing [47, 48].

Positive outcomes of music therapy have also been described for mood and behavioral symptoms such as agitation, depression, and anxiety [49], though the advantage of using music over other recreational activities is still debated [45].

Finally, improvements in quality of life are reported as a result of music-based interventions, both in patients and caregivers [44].

Notably, music can also serve as a risk prevention aid. Several lines of evidence suggest that lifetime musical activities, such as playing a musical instrument or dancing, are associated with a reduced risk of developing dementia [50]. Thus, music can also participate in enhancing cognitive reserve and delaying symptoms onset.

A severe limitation of the existing literature on music therapy is the paucity of details in the characterization of the sample (many articles refer broadly to dementia patients without specifying the exact diagnosis) and in the description of the protocols, which hinders reproducibility and confrontation across different studies. This shortcoming presumably accounts for many inconsistencies, and, due to these mixed results, there is still no agreement about several issues. For instance, there is no clear indication regarding optimal treatment duration, both with respect to the single session and to the total number of weeks needed for successful outcomes. According to a recent meta-analysis, treatments lasting less than 20 weeks provided better outcomes than longer ones [12]. Future studies are needed to assess whether the improvements found after treatment's end can last long-term.

6.4   Conclusions

In this chapter, we reviewed the existing literature related to music processing in the most common types of dementia. Overall, it appears that neurodegeneration leads to different selective impairment of musical skills depending on the clinical syndrome. Nonetheless, both anectodical and group studies consistently demonstrated that, even in the presence of cognitive and music processing deficits, some musical functions can still be preserved also in more advanced stages of the disease. Furthermore, music can be beneficial in managing psychological, cognitive, and behavioral symptoms in patients with dementia. Finally, the implementation of music-based interventions can be considered a promising tool for delaying symptoms of cognitive decline and for improving patients' quality of life in a relatively easy and cost-effective manner.

These findings, however, need to be expanded by more systematic studies. Literature in this field is often inconsistent, and it is particularly scant when considering less frequent forms of dementia. Contrasting results often originate between studies that employed different tasks and therapy protocols and that recruited subjects with heterogeneous musical backgrounds and disease duration. Thus, the lack of a consistent body of literature addressing all these variables prevents us from drawing solid conclusions. Future studies are also needed to develop specific batteries aimed at assessing spared musical functions in dementia, as well as questionnaires evaluating the individual history of music habits, in order to identify patients who will more likely benefit from a musical approach.

References

1. American Psychiatric Association D-TF. Diagnostic and statistical manual of mental disorders: DSM-5™. 5th ed. Charlottesville, VA: American Psychiatric Association; 2013.
2. Gauthier S, Rosa-Neto P, Morais JA, Webster C, editors. World alzheimer report 2021: journey through the diagnosis of dementia. London, England: Alzheimer's Disease International; 2021.
3. Erkkinen MG, Kim MO, Geschwind MD. Clinical neurology and epidemiology of the major neurodegenerative diseases. Cold Spring Harb Perspect Biol. 2018;10(4). 10.1101/cshperspect.a033118.
4. Rabinovici GD, Miller BL. Frontotemporal lobar degeneration: epidemiology, pathophysiology, diagnosis and management. CNS Drugs. 2010;24(5):375–98. https://doi.org/10.2165/11533100-000000000-00000.
5. Rascovsky K, Hodges JR, Knopman D, Mendez MF, Kramer JH, Neuhaus J, et al. Sensitivity of revised diagnostic criteria for the behavioural variant of frontotemporal dementia. Brain. 2011;134(Pt 9):2456–77. https://doi.org/10.1093/brain/awr179.
6. Gorno-Tempini ML, Hillis AE, Weintraub S, Kertesz A, Mendez M, Cappa SF, et al. Classification of primary progressive aphasia and its variants. Neurology. 2011;76(11):1006–14. https://doi.org/10.1212/WNL.0b013e31821103e6.
7. Elahi FM, Miller BL. A clinicopathological approach to the diagnosis of dementia. Nat Rev Neurol. 2017;13(8):457–76. https://doi.org/10.1038/nrneurol.2017.96.
8. Warren J. How does the brain process music? Clin Med (Lond). 2008;8(1):32–6. https://doi.org/10.7861/clinmedicine.8-1-32.
9. Särkämö T, Tervaniemi M, Huotilainen M. Music perception and cognition: development, neural basis, and rehabilitative use of music. Wiley Interdiscip Rev Cogn Sci. 2013;4(4):441–51. https://doi.org/10.1002/wcs.1237.
10. Koelsch S. Brain correlates of music-evoked emotions. Nat Rev Neurosci. 2014;15(3):170–80. https://doi.org/10.1038/nrn3666.
11. Sescousse G, Caldú X, Segura B, Dreher JC. Processing of primary and secondary rewards: a quantitative meta-analysis and review of human functional neuroimaging studies. Neurosci Biobehav Rev. 2013;37(4):681–96. https://doi.org/10.1016/j.neubiorev.2013.02.002.
12. Moreno-Morales C, Calero R, Moreno-Morales P, Pintado C. Music therapy in the treatment of dementia: a systematic review and meta-analysis. Front Med (Lausanne). 2020;7:160. https://doi.org/10.3389/fmed.2020.00160.
13. Hsieh S, Hornberger M, Piguet O, Hodges JR. Neural basis of music knowledge: evidence from the dementias. Brain. 2011;134(Pt 9):2523–34. https://doi.org/10.1093/brain/awr190.
14. Johnson JK, Chang CC, Brambati SM, Migliaccio R, Gorno-Tempini ML, Miller BL, et al. Music recognition in frontotemporal lobar degeneration and Alzheimer disease. Cogn Behav Neurol. 2011;24(2):74–84. https://doi.org/10.1097/WNN.0b013e31821de326.
15. Golden HL, Clark CN, Nicholas JM, Cohen MH, Slattery CF, Paterson RW, et al. Music perception in dementia. J Alzheimers Dis. 2017;55(3):933–49. https://doi.org/10.3233/jad-160359.
16. Goll JC, Kim LG, Hailstone JC, Lehmann M, Buckley A, Crutch SJ, et al. Auditory object cognition in dementia. Neuropsychologia. 2011;49(9):2755–65. https://doi.org/10.1016/j.neuropsychologia.2011.06.004.
17. Goll JC, Kim LG, Ridgway GR, Hailstone JC, Lehmann M, Buckley AH, et al. Impairments of auditory scene analysis in Alzheimer's disease. Brain. 2012;135(Pt 1):190–200. https://doi.org/10.1093/brain/awr260.
18. Clark CN, Golden HL, McCallion O, Nicholas JM, Cohen MH, Slattery CF, et al. Music models aberrant rule decoding and reward valuation in dementia. Soc Cogn Affect Neurosci. 2018;13(2):192–202. https://doi.org/10.1093/scan/nsx140.
19. Irish M, Piolino P. Impaired capacity for prospection in the dementias--theoretical and clinical implications. Br J Clin Psychol. 2016;55(1):49–68. https://doi.org/10.1111/bjc.12090.

20. Delbeuck X, Pollet M, Pasquier F, Bombois S, Moroni C. The clinical value of the faux pas test for diagnosing behavioral-variant frontotemporal dementia. J Geriatr Psychiatry Neurol. 2022;35(1):62–5. https://doi.org/10.1177/0891988720964253.
21. Van't Hooft JJ, Pijnenburg YAL, Sikkes SAM, Scheltens P, Spikman JM, Jaschke AC, et al. Frontotemporal dementia, music perception and social cognition share neurobiological circuits: a meta-analysis. Brain Cogn. 2021;148:105660. https://doi.org/10.1016/j.bandc.2020.105660.
22. Drapeau J, Gosselin N, Gagnon L, Peretz I, Lorrain D. Emotional recognition from face, voice, and music in dementia of the Alzheimer type. Ann N Y Acad Sci. 2009;1169:342–5. https://doi.org/10.1111/j.1749-6632.2009.04768.x.
23. Gagnon L, Peretz I, Fülöp T. Musical structural determinants of emotional judgments in dementia of the Alzheimer type. Neuropsychology. 2009;23(1):90–7. https://doi.org/10.1037/a0013790.
24. Arroyo-Anlló EM, Dauphin S, Fargeau MN, Ingrand P, Gil R. Music and emotion in Alzheimer's disease. Alzheimers Res Ther. 2019;11(1):69. https://doi.org/10.1186/s13195-019-0523-y.
25. Omar R, Hailstone JC, Warren JE, Crutch SJ, Warren JD. The cognitive organization of music knowledge: a clinical analysis. Brain. 2010;133(Pt 4):1200–13. https://doi.org/10.1093/brain/awp345.
26. Macoir J, Berubé-Lalancette S, Wilson MA, Laforce R, Hudon C, Gravel P, et al. When the wedding march becomes sad: semantic memory impairment for music in the semantic variant of primary progressive aphasia. Neurocase. 2016;22(6):486–95. https://doi.org/10.1080/13554794.2016.1257025.
27. Omar R, Henley SM, Bartlett JW, Hailstone JC, Gordon E, Sauter DA, et al. The structural neuroanatomy of music emotion recognition: evidence from frontotemporal lobar degeneration. NeuroImage. 2011;56(3):1814–21. https://doi.org/10.1016/j.neuroimage.2011.03.002.
28. Hsieh S, Hornberger M, Piguet O, Hodges JR. Brain correlates of musical and facial emotion recognition: evidence from the dementias. Neuropsychologia. 2012;50(8):1814–22. https://doi.org/10.1016/j.neuropsychologia.2012.04.006.
29. Downey LE, Blezat A, Nicholas J, Omar R, Golden HL, Mahoney CJ, et al. Mentalising music in frontotemporal dementia. Cortex. 2013;49(7):1844–55. https://doi.org/10.1016/j.cortex.2012.09.011.
30. Barquero S, Gomez-Tortosa E, Baron M, Rabano A, Munoz DG, Jimenez-Escrig A. Amusia as an early manifestation of frontotemporal dementia caused by a novel progranulin mutation. J Neurol. 2010;257(3):475–7. https://doi.org/10.1007/s00415-009-5393-y.
31. Baird A, Samson S. Memory for music in Alzheimer's disease: unforgettable? Neuropsychol Rev. 2009;19(1):85–101. https://doi.org/10.1007/s11065-009-9085-2.
32. Beatty WW, Winn P, Adams RL, Allen EW, Wilson DA, Prince JR, et al. Preserved cognitive skills in dementia of the Alzheimer type. Arch Neurol. 1994;51(10):1040–6. https://doi.org/10.1001/archneur.1994.00540220088018.
33. Cowles A, Beatty WW, Nixon SJ, Lutz LJ, Paulk J, Paulk K, et al. Musical skill in dementia: a violinist presumed to have Alzheimer's disease learns to play a new song. Neurocase. 2003;9(6):493–503. https://doi.org/10.1076/neur.9.6.493.29378.
34. Cuddy LL, Duffin JM, Gill SS, Brown CL, Sikka R, Vanstone AD. Memory for melodies and lyrics in Alzheimer's disease. Music Percept. 2012;29(5):479–91.
35. Cuddy LL, Sikka R, Vanstone A. Preservation of musical memory and engagement in healthy aging and Alzheimer's disease. Ann N Y Acad Sci. 2015;1337:223–31. https://doi.org/10.1111/nyas.12617.
36. Ménard MC, Belleville S. Musical and verbal memory in Alzheimer's disease: a study of long-term and short-term memory. Brain Cogn. 2009;71(1):38–45. https://doi.org/10.1016/j.bandc.2009.03.008.
37. Baird A, Brancatisano O, Gelding R, Thompson WF. Characterization of music and photograph evoked autobiographical memories in people with Alzheimer's disease. J Alzheimers Dis. 2018;66(2):693–706. https://doi.org/10.3233/jad-180627.

38. Baird A, Brancatisano O, Gelding R, Thompson WF. Music evoked autobiographical memories in people with behavioural variant frontotemporal dementia. Memory. 2020;28(3):323–36. https://doi.org/10.1080/09658211.2020.1713379.
39. Baird A, Umbach H, Thompson WF. A nonmusician with severe Alzheimer's dementia learns a new song. Neurocase. 2017;23(1):36–40. https://doi.org/10.1080/13554794.2017.1287278.
40. Cho H, Chin J, Suh MK, Kim HJ, Kim YJ, Ye BS, et al. Postmorbid learning of saxophone playing in a patient with frontotemporal dementia. Neurocase. 2015;21(6):767–72. https://doi.org/10.1080/13554794.2014.992915.
41. Miller BL, Boone K, Cummings JL, Read SL, Mishkin F. Functional correlates of musical and visual ability in frontotemporal dementia. Br J Psychiatry. 2000;176:458–63. https://doi.org/10.1192/bjp.176.5.458.
42. Fletcher PD, Downey LE, Witoonpanich P, Warren JD. The brain basis of musicophilia: evidence from frontotemporal lobar degeneration. Front Psychol. 2013;4:347. https://doi.org/10.3389/fpsyg.2013.00347.
43. Brancatisano O, Baird A, Thompson WF. A 'music, mind and movement' program for people with dementia: initial evidence of improved cognition. Front Psychol. 2019;10:1435. https://doi.org/10.3389/fpsyg.2019.01435.
44. Särkämö T, Tervaniemi M, Laitinen S, Numminen A, Kurki M, Johnson JK, et al. Cognitive, emotional, and social benefits of regular musical activities in early dementia: randomized controlled study. Gerontologist. 2014;54(4):634–50. https://doi.org/10.1093/geront/gnt100.
45. Sánchez A, Maseda A, Marante-Moar MP, de Labra C, Lorenzo-López L, Millán-Calenti JC. Comparing the effects of multisensory stimulation and individualized music sessions on elderly people with severe dementia: a randomized controlled trial. J Alzheimers Dis. 2016;52(1):303–15. https://doi.org/10.3233/jad-151150.
46. Moussard A, Bigand E, Belleville S, Peretz I. Learning sung lyrics aids retention in normal ageing and Alzheimer's disease. Neuropsychol Rehabil. 2014;24(6):894–917. https://doi.org/10.1080/09602011.2014.917982.
47. Racette A, Bard C, Peretz I. Making non-fluent aphasics speak: sing along! Brain. 2006;129(Pt 10):2571–84. https://doi.org/10.1093/brain/awl250.
48. Warren JD, Warren JE, Fox NC, Warrington EK. Nothing to say, something to sing: primary progressive dynamic aphasia. Neurocase. 2003;9(2):140–55. https://doi.org/10.1076/neur.9.2.140.15068.
49. Lam HL, Li WTV, Laher I. Effects of music therapy on patients with dementia-a systematic review. Geriatrics (Basel). 2020;5(4) https://doi.org/10.3390/geriatrics5040062.
50. Verghese J, Lipton RB, Katz MJ, Hall CB, Derby CA, Kuslansky G, et al. Leisure activities and the risk of dementia in the elderly. N Engl J Med. 2003;348(25):2508–16. https://doi.org/10.1056/NEJMoa022252.

Parkinson's Disease Treatment: The Role of Music Therapy

7

Livio Claudio Bressan

7.1 Introduction

In the last decades, Parkinson's disease management approach has changed, becoming both global and person-centered at the same time and ready to focus both on movement disorders and on the sick person in its wholeness.

On these bases, therapy will succeed not only in treating the "cardinal" motor symptoms but also in balancing all the other spheres of everyday and social patient's life. In other words, the whole quality of life, both actual and personal, should become the main principle of efficiency and primary outcome beneath clinical trials [1].

The framework of the therapeutic outcome is expanded: education and training, diet, physiotherapies, and finally alternative therapies are all part of the global program [2]. However, to discuss the specific role of music therapy for PD, we should first introduce the complementary therapies, which include music therapy, and the new trend that excludes the word "therapy" from complementary methods if they're not strictly medical, using instead the word "well-being."

7.2 Complimentary Methods

In Western nations, more and more people decide to utilize "complimentary or unconventional medicine" (also called "alternative") to treat both chronic disorders like headaches, backaches, anxious-depressive neurosis, and in some cases looking for hope in incurable or terminal illness. However, this field is quite hard to define: doctors and other healthcare operators with perfect curriculum act within it, but also some ambiguous healers do, and all of them have in common the adoption of

L. C. Bressan (✉)
Bassini Hospital, University of Milan, Milan, Italy

B. Colombo (ed.), *The Musical Neurons*, Neurocultural Health and Wellbeing, https://doi.org/10.1007/978-3-031-08132-3_7

therapies that are not taught at medical school or available in hospitals. There's a lot of debate about those methods between supporters and detractors and I'd rather abstain from taking part to it, but, in brief, complementary medicines promote healthcare by scanning every aspect: physical, psychological, emotional, cognitive, social, and spiritual.

The emphasis is put on word "holistic," from Greek "Olos" meaning "all," and it never omits, but it enhances the emotional sphere and the personal feeling of the sick person. In this concept, illness doesn't come only from biological dysfunctions, but it also concerns the whole person even with its deep, psychological, and social aspects. In the United States and in the rest of Europe, holistic interventions are more common than in Italy and are led by the practitioner himself in agreement with doctors, nurses, and other health operators. According to World Health Organization (WHO), 15% of Italians turn to alternative medicine. In this context, the Technical Scientific Committee of Lombardy Region (CTS) assumed that supporting new therapeutic techniques is "ethically unacceptable" until (1) their safety and effectiveness aren't at least at the same level of the other available therapies, and (2) provided safety and effectiveness criteria, in order to protect both citizens and operators themselves. Among alternative methods, music therapy plays an important role since it is widely adopted in medical and nursing environments and it has gained mass media attention.

7.3   Music Therapy (MT)

According to the World Federation of Music Therapy (WFMT)'s definition, "MT is the professional use of music and its components as an intervention in medical, educational, and everyday environments with individuals, groups, families, or communities who seek to optimize their quality of life and improve their physical, social, communicative, emotional, intellectual, and spiritual health and wellbeing".

However, today the use of MT in the medical environment is random and depends on the medical directors' culture and purposes. Besides cultural reasons, one of the critical points is the lack of regulation about the music therapist's curriculum. In order to find a quick answer to the question, the experts advise the nurses and the other health operators to increase their professional skills with a basic knowledge of MT. Since MT could be useful for all ages and in different healthcare areas, music therapist has to be able to work with children and adults with emotional, physical, mental, or psychological problems.

The sound/musical component applies as a therapy, especially in two operational areas: psychotherapy and rehabilitation. In the first area, music becomes a nonverbal communication channel enhancing interpersonal relation, while in rehabilitative area, the rhythmic component acts like a stimulus for motor, vocal, or cognitive function. The sound/musical stuff choice depends on the music therapy intervention requirements, on its modes, and on its final users. There are two music therapy modes: receptive and active. In the receptive MT, the subject is in listening mode, involving not only his auditory canal but also his full body, living both physical and emotional experience. On the contrary, in active MT, the subject takes part individually or in group by producing music with his voice, body, musical instruments, or dancing being stimulated by rhythm and melody.

Music is recommended as therapy for different diseases since it could also help noncollaborative patients: bedridden, in a coma state, or if both affected with cognitive degeneration and in development but in mental retardation. Back to the objective of this article, we created a set of rehabilitative interventions based on harmonic integration between scientific medicine and alternative medicine, aiming to soothe individuals with Parkinson Disease's (PD) pain, not to unfairly promise them an impossible healing but to agree with them in order to avoid both isolation and helplessness mood, frequently gripping patients and their relatives.

7.4   Rhythm Therapy for Individuals with PD

It's easy to explain: please, hands up who has never realized he was tapping his foot while listening to music? However, this rhythmic movement is two times synchronized, since it has the same passage of the stimulus and stimulus and its reaction occur in the same moment. Actually to produce synchronization between the foot tap and a regular rhythm, an anticipation system at a cerebral level should work; in fact the synchronization is different from the reaction caused in answering the phone. The answering signal of synchronization isn't the rhythmic stimulus but the time interval among the following beats and, if the frequency is regular, synchronization shows by anticipating the following beat. Since synchronization is an irrepressible phenomenon, we may use rhythm by inducing a movement in a part of the body and then accelerating or slowing the frequency. Rehabilitation through rhythm could be extended to voice, because phonation also depends on a motor activity. Besides, rhythmical experience has a natural "social feature" in dance, march, and choral singing, which are typical rhythmical collective phenomena. On this basis, it comes out that the auditive system is a very fast processing system for acoustic information and an excellent decoding system for the rhythmical aspect of the sensorial signal.

Michael Thaut, from *Center for Neurorehabilitation Research of Colorado State University*, carried out several studies on the effect of auditive rhythmic on motor performance, achieving to claim that individuals with PD don't need ganglion integrity to take advantage of the rhythmical-auditive facilitation [3].

Physiology's studies hypothesize the existence of an auditive-motor path through which sound, using reticular spinal connectors, activates spinal motoneuron activity [4, 5]. However, it's not clear if some specific neural structures are responsible for rhythmic motor synchronization or auditive stimulus is *tout court* directly projected in the motor system circuits.

Masahiro demonstrated how rhythm could stimulate accurate and fast motor answers even under the conscious perception threshold [6]. Finally, Sutteerawattananon claims the auditive input influences the *Secondary Motor Area*, acting like a stimulus to reduce akinesia and bradykinesia [7].

In our team [8], we utilize specific exercises associated with rhythm as a fundamental pre-requirement to promote the patients' motor "decision-making." Admittedly, the difficulty in acting simultaneous movements and programmed strategies to reach their goals is well known in PD patients. In a recent study, still going on, we selected 20 patients with idiopathic PD, randomized into two groups of 10 patients. The first group was treated with traditional rehabilitation (without rhythm). The second group was assigned to rhythm therapy. None of the patients have modified their pharmacological therapy during this study, and none showed a relevant cognitive deficit. Evaluation was carried out before treatment, at the sixth session, and at the end of the study using the UPDRS scale, the LSI scale (Life Satisfaction Index), and the Walking Test. For the statistical analysis, we used Student's t-test. Values of $p < 0.05$ were considered significant with a confidence interval 95%. **Even if the number of patients enrolled is low and final data are not published yet, first results could confirm the rehabilitative effect of rhythm therapy on PD disease. Results are favorable, e**specially in patients with more severe symptoms compared to the less serious ones. In conclusion, rhythmical motor induction could *increase* speed and precision in the movement execution. Moreover, individuals with PD may take advantage of rhythm also during everyday life by imaging the learned rhythmic exercise. Finally, the rhythmical experience with its social and ludic character showed a positive influence on individuals with PD quality of life.

7.5   Music Therapy for Parkinson's Disease

Parkinson's disease can be evaluated from various points of view: music therapy could be a complementary therapy option. Many patients affected by Parkinson's disease have experienced different music therapy programs focused on the motor component, which can be addressed through listening, body rhythm, and rhythmic auditory stimulation. In order to improve quality of life, other music therapy programs focused on communication, swallowing, breathing, and emotional aspects were standardized through exercises focusing on singing, either individually or in groups. We concluded that music therapy programs can achieve improvements in various areas of patients with Parkinson's disease. Music therapy has been defined as the use of sounds and music within an evolving patient–therapist relationship to support and develop physical, mental, and social spiritual well-being [9]. Music has cognitive, psychosocial, behavioral, and motor benefits for patients affected by neurological disorders such as Parkinson's disease or dementia [10, 11]. The value that music offers both to human health and well-being provides a framework for the development of non-pharmaceutical treatments for neurological disorders [12]. In this context, making music is a powerful way to engage a multisensory and motor network. These multimodal effects on emotional brain areas can be used to facilitate the rehabilitation of neurological disorders [13]. A recent study confirmed that a music-based physical therapy program is able to improve functional mobility in patients with PD [14]. According to De Bartolo et al. [15], in Parkinson's disease, the timing of repetitive sequences of internally generated automatic movements is compromised. The consequence of this deficit is on the alteration of gait patterns, including shorter steps, slower gait, loss of rhythm, and trunk instability. In addition to cognitive and emotional disturbances, the most frequent non-motor symptoms are autonomic, sensory, gastrointestinal dysfunctions, and sleep disorders [16]. Music therapy and other rhythm-based interventions can offer benefits for patients with Parkinson's disease and other related movement disorders [17]. According to De Luca et al. [18], music-based rehabilitation for gait training can be considered a powerful strategy. Thaut et al. described that rhythmic auditory cues are able to improve gait and motor behaviors in Parkinson's disease [19]. A systematic review and meta-analysis of 17 studies concerning music therapy and Parkinson's disease, with a total sample of 598 participants, concluded that music therapy improves motor function, balance, freezing of gait, gait speed, and mental health in patients with Parkinson's disease [20]. Furthermore, a group singing program with deep breathing training and song learning can promote memory, language, executive function, and respiratory muscle strength in adults with Parkinson's disease [21]. Choral singing has also been studied to help patients affected with Parkinson's disease with social isolation and mood disturbances [22]. In conclusion, music therapy in Parkinson's disease has a scientific basis to support physical, emotional, and social benefits [23]. Some recent work (such as the paper of Barnish and Barran) [24] introduced performing arts and music therapy, considering their possible positive effects on Parkinson's disease. The favorable effects of music therapy programs on different spheres of human development in Parkinson's patients were confirmed. The majority of studies focused on the motor component, stimulating it through programs based on listening music, body rhythm, and rhythmic auditory stimulation. Group singing, optimizing the quality of life, and listening to the music help maintain or improve the cognitive functions of PD patients.

7.6 Holistic Musical Techniques

On this basis, we should notice that some years ago Lombardy Region in Italy banned the use of the term "therapy" if referring to not strictly medical professions. Meanwhile, the role of Bio-natural Disciplines Operators was defined, referring to "activities and practices focused in maintaining or recover wellbeing." These practices (which are not sanitary) enhance the individual vital energy through natural methods whose effectiveness was certified in the cultural and geographical environment where these special disciplines were developed.

In accordance with the Regional Law 2/2005, Bio-Natural Disciplines Technical Scientific Committee (CTS-DBN) was established. It is based on the Education Institution and Operators Association promoting and broadcasting Bio-Natural Disciplines culture and methods across the Lombardy region and the whole country. Many represented that disciplines have been carrying out important research activities, cooperating both with hospitals and universities in the Lombardy Region. For this reason, they become best practices also at the national level.

The author, neurologist, and musician, in line with CTS-DBN statements, created the educational program Holistic Musical Techniques, a bio-natural discipline using music (sound, rhythm, melody, and harmony) and musical instruments to perform a process in order to maintain or recover well-being, stimulating vital resources. Holistic Musical Techniques are based on a scientifically proven operating model, for example, the Argentine Benenzon's model based on music and movement to encourage socializing. Moreover, an insight is derived from Francesco Padrini's work. It was a psychologist, psychotherapist, and sexologist invented Bioenergetic Massage® and Grounding Massage® techniques and founded Integrated Psychophysiognomy Society – face-body-character (SPI). All these techniques are well integrated with active and receptive bio music therapy. Finally, the author's method, based on holistic musical, artistic, and physical techniques, is focused on maintaining or recovering psychophysical elderly people's health. After a 3-year program, including 850 hours of on-site lessons, internship, thesis preparation, and final examination, the graduated operator will be certified in the Lombardy region Bio-natural Disciplines Operators Register. Consequently, it will be able to support its patients in customized bio-musical programs, scheduled for all different ages. In conclusion, TMOs belong to Bio-natural Disciplines: using the auditive canal, they involve the whole person. We could say *"music matters too."*

7.7 Conclusions

The most relevant improvement in PD treatment in the last years should be credited to the more and more clear and widespread awareness of how worthy the new disease treatment philosophy is. In fact, it's global, customized, and patient centered, rather than focusing exclusively on motor diseases. This change of perspective's influences is several and deep. The primary treatment target is the patient, rather than the motor autonomy. The treatment is tailored as an integrated and customized program, rather than only active principles drugging. Awareness toward patients and their relatives is focused on a real therapeutic alliance rather than just compliance. For these reasons, music therapy and TMOs are very well accepted among alternative and bio-natural disciplines with the aim of maintaining or recovering well-being. As a matter of fact, TMOs are not intentionally considered medical treatment: they positively tend to encourage vital resources of individuals through natural stimuli like music, rhythm, and natural sounds in order to promote well-being and health.

References

1. Jankovic J. New and emerging therapies for Parkinson's disease. Arch Neurol. 1999;56:785–90.
2. Manyam BV, Sanchez-Ramos JR. Traditional and complementary therapies for Parkinson' s disease. Adv Neurol. 1999;80:565–74.
3. Thaut MH, Ruth RR, Thenille BJ, Corene PH, McIntosh GC. Rhythmicauditory stimulation for reduction of falls in Parkinson's disease: a randomized controlled study. Clin Rehabil. 2019;33(1):34–43.

4. Rossignol S, Melvill JG. Audiospinal influences in man studied by the H-reflex and its possible role in rhythmic movement synchronized to sound. Electroencephalogr Clin Neurophysiol. 1976;41:83–92.
5. Tecchio F, Salustri C, Thaut MH, Pasqualetti P, Rossini PM. Conscious and unconscious adaptation: a MEG study of cerebral responses to rhythmic auditory stimuli. Exp Brain Res. 2000;135:222–30.
6. Masahiro O, Masahiro S, Kazutoshi K. Paired synchronous rhythmic finger tapping without an external timing Cue shows greater speed increases relative to those for solo tapping. Nature. 2017;7:43987.
7. Sutteerawattananon M, Morris GS, Etnyre BR, Jankovich J, Protas EJ. Effects of visual and auditory cues on gait in individuals with Parkinson's disease. J Neurol Sci. 2004;219(1–2):63–9.
8. Bergna A, Zanfagna E, Ballabio A, Vendramini A, Bressan LC, Solimene U. Efficacy of the osteopathic treatment in Parkinson's disease. J Bull Reabilitat Med. 2021;20:2.
9. Bunt L, Hoskyns S, Swami S. The handbook of music therapy. London: Routledge; 2013.
10. Dowson B, Schneider J. Online singing groups for people with dementia: scoping review. Public Health. 2021;194:196–201.
11. Fodor DM, Breda XM, Valean D, Marta MM, Perju-Dumbrava L. Music as add on therapy in the rehabilitation program of Parkinson's disease patients: a romanian pilot study. Brain Sci. 2021;11:569.
12. Brancatisano O, Baird A, Thompson WF. Why is music therapeutic for neurological disorders? The therapeutic music capacities model. Neurosci Biobehav Rev. 2020;112:600–15.
13. Altenmüller E, Schlaug G. Neurologic music therapy: the beneficial effects of music making on neurorehabilitation. Acoust Sci Technol. 2013;34:5–12.
14. Da Silva LK, Brito TSS, de Souza LAPS, Luvizutto GJ. Music-based physical therapy in Parkinson's disease: an approach based on international classification of functioning, disability and health. J Bodyword Mov Ther. 2021;26:524–9.
15. De Bartolo D, Morone G, Giordani G, Antonucci G, Russo V, Fusco A, Marinozzi F, Bini F, Spitoni GF, Paolucci S, et al. Effect of different music genres on gait patterns in Parkinson's disease. Neurol Sci. 2020;41:575–82.
16. Massano J, Bhatia KP. Clinical approach to Parkinson's disease: features, diagnosis, and principles of management. Cold Spring Harb Perspect Med. 2012;2:8870.
17. Devlin K, Alshaikh JT, Pantelyat A. Music therapy and music-based interventions for movement disorders. Curr Neurol Neurosci Rep. 2019;19:83.
18. De Luca R, Latella D, Maggio MG, Leonardi S, Sorbera C, Di Lorenzo G, Balletta T, Cannavò A, Naro A, Impellizzeri F, et al. Do PD patients benefit from music-assisted therapy plus treadmill-based gait training? An exploratory study focusing on behavioral outcomes. Int J Neurosci. 2020;130:933–40.
19. Braunlich K, Seger CA, Jentink KG, Buard I, Kluger BM, Thaut MH. Rhythmic auditory signals shape neural network recruitment in Parkinson's disease during repetitive motor behavior. Eur J Neurosci. 2019;49:849–58.
20. Zhou Z, Zhou R, Wei W, Luan R, Li K. Effects of music-based movement therapy on motor function, balance, gait, mental health, and quality of life for patients with parkinson's disease: a systematic review and meta-analysis. Clin Rehabil. 2021;35:026–921.
21. Barnish J, Atkinson RA, Barran SM, Barnish MS. Potential benefit of singing for people with parkinson's disease: a systematic review. J Parkinsons Dis. 2016;6:473–84.
22. Fogg-Rogers L, Buetow S, Talmage A, McCann CM, Leão SH, Tippett L, Leung J, McPherson KM, Purdy SC. Choral singing therapy following stroke or parkinsons disease: an exploration of participants experiences. Disabil Rehabil. 2016;38:952–62.
23. Westheimer O, Mcrae C, Henchcliffe C, Fesharaki A, Glazman S, Ene H, Bodis-Wollner I. Dance for PD: a preliminary investigation of effects on motor function and quality of life among persons with Parkinson's disease. J Neural Transm. 2015;122:1263–70.
24. Barnish MS, Barran SM. A systematic review of active group-based dance, singing, music therapy and theatrical interventions for quality of life, functional communication, speech, motor function and cognitive status in people with Parkinson's disease. BMC Neurol. 2020;20:371.

It Thrills My Soul to Hear the Songs: The Case of Musicolepsia

8

Arturo Nuara

> *Some men there are, love not a gaping pig;*
> *Some that are mad, if they behold a cat;*
> *And others, when the bagpipe sings i' the nose,*
> *Cannot contain their urine.*
> *Shakespeare, Merchant of Venice (Act IV, Scene I).*

8.1   Introduction

The term "musicogenic epilepsy" (or "musicolepsia") was first adopted by Critchley [1] to classify a series of neurological cases characterized by *"the occurrence of epileptiform attacks in factual association with the hearing of music."*

Musicogenic seizures (MSs) are a rare phenomenon, showing a prevalence as low as 1 case per 10,000,000 population [2]. However, this prevalence may be underestimated due to the high latency between stimulus onset and seizure beginning, as well as to the limited musical education of people and physicians [3].

Musicolepsia is usually enclosed in the framework of reflex epilepsies, even if not all cases fulfill the criteria for such a classification. In fact, in most patients, music is not the only stimulus that evokes seizures, as spontaneously

A. Nuara (✉)
Consiglio Nazionale delle Ricerche, Istituto di Neuroscienze, Parma, Italy

© The Author(s), under exclusive license to Springer Nature
Switzerland AG 2022
B. Colombo (ed.), *The Musical Neurons*, Neurocultural Health and Wellbeing,
https://doi.org/10.1007/978-3-031-08132-3_8

occurring seizures and music-associated ones often coexist. In addition, different features of the musical stimulus can trigger the seizure, spanning from a specific genre to musical instruments, to the emotional charge of music; rarely, seizures are selectively provoked by specific tunes [4] or songs [5]. Finally, the sporadic cases where the musical stimulus is highly specific, e.g., a particular frequency band of sound [6], are unlikely categorizable as musicolepsia, since the ictogenic stimulus is not endowed with the complexity needed to define music for itself.

When we listen to a musical piece, in addition to the auditory cortices, widespread brain networks are recruited, including frontotemporal circuits involved in reward and expectance [7], networks for the syntactic organization of sounds [8], and motor areas involved in rhythmic entrainment [9]. Moreover, listening to intensely pleasurable music additionally activates limbic regions, subcallosal, orbitofrontal, and frontal polar cortex [10], indicating the presence of a pervasive brain activity when listening to "emotionally charged" music.

Given this evidence, the heterogeneity in terms of epileptogenic trigger (e.g., rhythmic features, melodic contour, emotional charge) may be justified as a consequence of a hyperexcitability confined to particular brain regions underlying a specific domain of music perception. Thus, offering the possibility to link clinical symptoms with specific brain topographies, musicolepsia may provide relevant insight into the functional role of brain regions in multiple domains of music processing.

Adopting the same systematic approach described in Nuara et al. [11], I upgraded the review of the cases of musicogenic epilepsy described so far in the neurological literature (last record: November 2021), finally including 73 studies (56 single case report, 11 series of case reports, 1 retrospective study, 4 monograph chapters), from which 146 clinical cases have been extracted (the recorded cases are extensively reported in Table 8.1 and summarized in Fig. 8.1). The findings will be discussed, giving special emphasis to the role of the emotional charge of music on ictogenesis and its putative brain substrates.

Table 8.1 Clinical and neurophysiological features of the reviewed cases of musicogenic epilepsy

ID	Authors	Age, sex	Music training	onset (age)	Music genre	Emotional component	Other seizures	Latency	Aura type
1	Merzheevsky, 1884 (cited in [1])	–	–	–	Unfamiliar tune	–	No	–	Dizziness
2	Steinbrugge, 1889 (cited in [1])	45,M	–	15	Unspecific	–	No	–	–
3	Trutovski, 1892 (cited in [1])	31,M	–	–	Piano music	–	–	–	–
4	Oppenheim, 1905 (cited in [1])	45,F	–	–	Unspecific	Yes	–	–	Anxiety
5	Katchatchoff, 1907 (cited in Vanelle [12])	–,M	–	–	Specific orchestral music	–	–	–	–
6	Lwoff, 1913 (cited in Vanelle [12])	55,F	–	–	Specific song ("La Marseillaise")	Yes	–	–	–
7	Bechterew, 1914 (cited in [1])	–,M	–	–	Unspecific	Yes	No	–	–
8	Marchand, 1926 (cited in Vanelle [12])	50,F	–	–	Noises, music, songs	–	–	–	–
9	Redlich, 1929 (cited in [1])	–,–	–	–	Violin music in particular	–	–	–	–
10	Goldstein, 1932 (cited in [1])	32,M	–	27	Unspecific	Yes	–	–	–

(continued)

Table 8.1 (continued)

ID	Authors	Age, sex	Music training	onset (age)	Music genre	Emotional component	Other seizures	Latency	Aura type
11	Bonhoeffer, 1932 (cited in [1])	-,F	–	30	Unspecific	–	–	–	–
12	Nikitin, 1935 (cited in [1])	30,M	–	23	A specific song or its imagination	–	–	–	Right sided paresthesia
13	[1]	25,F	–	17	Piano, organ, classical. Not dance music	No	No	–	–
14	[1]	46,F	–	42	Sad music. Not dance music.	Yes	Deep-conversation induced seizures	–	–
15	[1]	51,F	–	44	Dance music, organ. Well-punctuated rhythm. Sing-song voices.	No	No	–	–
16	[1]	33,F	–	30	Classical music, waltzes	Yes	No	–	–
17	[1]	22,M	Pianist	–	Playing piano	No	–	–	–
18	[1]	26,M	Musician	22	Playing piano and cello	No	Yes (unspecified)	–	–
19	[1]	62,M	–	32	Organ, piano, opera	No	–	–	–
20	[1]	25,M	No	–	Brass playing bass notes	No	–	–	–

			Pianist				"Thinking out tune" induced seizure			
21	[1]	31,F	Pianist	25	Playing piano	–	"Thinking out tune" induced seizure	–	–	–
22	[1]	40,M	–	32	Concerts	No	–	–	–	Auditory aura in left ear
23	[1]	-,M	–	–	Piano	–	–	–	–	Psychological aura
24	Taylor, 1942 (cited in Vanelle [12])	20,M	–	–	Unspecific music, referee's whistle	–	–	–	–	–
25	Critchley, 1942 (cited in Vanelle [12])	22,M	–	–	Piano, organ	Yes	–	–	–	–
26	Critchley, 1942 (cited in Vanelle [12])	30,M	–	–	Specific song	–	–	–	–	–
27	[13]	44,F	–	–	Unspecific	Yes	–	–	–	–
28	Reese, 1948 (cited in Vanelle [12])	42,F	–	–	Coral or sentimental music	Yes	–	–	–	–
29	Reese, 1948 (cited in Vanelle [12])	57,F	–	–	Music, singing	–	–	–	–	–
30	Hamoir and Titeca, 1948 (cited in Joint et al. [14])	20,F	–	–	Sad songs, where violin predominated	Yes	Generalized T-C seizure	–	–	Epigastric distress

(continued)

Table 8.1 (continued)

ID	Authors	Age, sex	Music training	onset (age)	Music genre	Emotional component	Other seizures	Latency	Aura type
31	Stubbe-Teglbjaerg [15]	62,F	-	54	Classical music	No	-	-	-
32	Stubbe-Teglbjaerg [15]	38,M	-	22	Classical music	Yes	Non-musicogenic fits	-	-
33	Stubbe-Teglbjaerg [15]	36,F	-	28	Unspecific	No	Music-related only 2 years after onset	-	-
34	Stubbe-Teglbjaerg [15]	54,M	-	37	Classical music	Yes	Remembering-music induced	-	Pleasure sensation
35	Arellano et al. [16]	24,F	-	-	Sounds of specific frequency	-	-	-	-
36	Dow [17]	29,F	-	-	Mornful music	Yes	-	-	-
37	Krayenbuhl, 1952 (cited in Vanelle [12])	37,F	-	-	Unspecific	-	-	-	-
38	Vercelletto [18]	38,F	-	-	Unspecific	Yes	-	-	-
39	Hoheisel and Walch [19]	24,M	-	-	Sax, trumpet, sirens and whistles	-	-	-	-
40	Levi [20]	38,M	-	-	Unspecific	-	-	-	-
41	Szbor, 1955 (cited in Vanelle [12])	32,F	-	-	Unspecific	Yes	-	-	-
42	Bickford [21]	23,F	-	-	Exotic jazz music	-	-	-	-

43	Weber [22]	45,M	–	–	Unspecific, organ music in particular	–	–	–	–
44	Weber [22]	23,F	–	–	Slow, romantic, popular waltz music	–	–	–	–
45	Hess, 1956 (cited in Weber [22])	50,M	–	–	Unspecific	–	–	–	–
46	Daly and Barry [23]	24,F	–	16	Jazz music	No	Non-musicogenic fits	–	Fear (unconstant)
47	Daly and Barry [23]	41,M	–	24	Two popular songs	No	Episodes od deja-vu and TC seizures	–	No
48	Daly and Barry [23]	26,M	Pianist	3	Playin piano	Yes	Generalized nocturnal fits	–	No
49	Cvetko, 1957 (cited in Vanelle [12])	23,F	–	–	Unspecific	–	–	–	–
50	Cvetko, 1957 (cited in Vanelle [12])	51,F	–	–	Unspecific	–	–	–	–
51	Cvetko, 1957 (cited in Vanelle [12])	28,M	–	–	Unspecific	–	–	–	–
52	Reifenberg [24]	46,F	–	–	Music and knocking noises	–	–	–	–

(continued)

Table 8.1 (continued)

ID	Authors	Age, sex	Music training	onset (age)	Music genre	Emotional component	Other seizures	Latency	Aura type
53	Barrios del Risco and Esslen [25]	44,F	–	–	Popular music	Yes	–	–	–
54	Anastasopulos, 1958 (cited in Titeca [26])	22,M	–	–	Unspecific	–	–	–	–
55	Anastasopulos, 1958 (cited in Titeca [26])	27,M	–	–	Unspecific	–	–	–	–
56	Anastasopulos, 1958 (cited in Titeca [26])	45,F	–	–	Unspecific	–	–	–	–
57	Piotrowski [27]	-,F	–	–	Sentimental songs by female artists	Yes	–	–	–
58	Bash and Bash-Lietchti [28]	45,F	–	–	Sentimental music	Yes	–	–	–
59	Lennox, 1960 (Cited in Titeca [26])	-,F	–	–	Unspecific	–	No	–	Panic
60	Lennox, 1960 (Cited in Titeca [26])	69,F	–	25	Piano, vocal, instrumental classical or jazz	–	Non-musicogenic seizures later in life	–	–
61	Lennox, 1960 (Cited in Titeca [26])	-,M	–	–	Unspecific	–	No	–	Unpleasant visceral sensation, paraesthesiae

							Non-musicogenic fits		Sense of impending consciousness
62	Poskanzer [6]	62,M	No	56	Church bell	No	–	17''	Sense of impending consciousness
63	Joynt [14]	62,F	No	56	Organ sound	No	Nocturnal generalized seizures	441''	Anxiousness
64	Yvonneau and Barros-Ferreira [29]	72,F	–	–	Loud and unexpected sounds	No	Audiogenic (loud sound)	–	–
65	Gormik et al. [30]	32,M	–	–	Unspecific	–	–	–	–
66	Gormik et al. [30]	33,M	–	–	Unspecific	–	–	–	–
67	Perpiniotis [31]	16,F	–	–	Popular sentimental song	Yes	–	–	–
68	Forster [32]	-,M	–	–	Orchestral music	–	–	–	–
69	Toivakka and Lehtinen [33]	31,M	No	26	Sentimental vocal and romantic popular music	Yes	Nocturnal generalized seizures	–	Feeling of happiness
70	Titeca [26]	38,F	–	–	Languid tune	–	–	–	–
71	Dearman [34]	38,F	No	–	Selected songs	Yes	Yes (unspecified)	–	–
72	Strang et al., 1966 (cited in Vanelle [12])	43,M	–	–	Playing classical music	–	–	–	–

(continued)

Table 8.1 (continued)

ID	Authors	Age, sex	Music training	onset (age)	Music genre	Emotional component	Other seizures	Latency	Aura type
73	Gastaut and Tassinari [35]	16,F	–	–	Sentimental music	Yes	No	–	–
74	Shaper et al., 1967 (cited in Vanelle [12])	28,M	–	–	Popular songs	–	–	–	–
75	Vizioli et al. [36]	35,F	–	31	Popular Sardinian songs	Yes	–	–	–
76	Fujinawa [4]	70,F	–	–	Specific song	Yes	–	–	–
77	Mundt, 1980 (cited in Vanelle [12])	25,F	–	–	Popular songs	–	–	–	–
78	Newman and Saunders [37]	39,F	–	–	Light and background music	No	Nocturnal generalized seizures	–	Feeling anxious and sweaty
79	Sutherling [38]	67,M	No	–	Playing hymn with organ	No	Yes (unspecified)	–	–
80	Vanelle [12]	44,M	–	–	Classical music	–	–	–	–
81	Vanelle [12]	33,M	–	–	Popular repetitive music	–	–	–	–
82	Brien and Murray [39]	53,F	No	15	Particular singer's voices	–	–	Seconds/minutes	Epigastric sensation, fear and depersonalization
83	Byun et al. [40]	40,F	–	–	Popular korean songs	–	–	1'30''	–
84	Jallon et al. [41]	–,M	–	–	Music with high emotional charge	Yes	–	–	–
85	Smeijsters and van den Berk [42]	40,F	Clarinet		Unspecific	Yes	–	–	–

No.	Reference	Age, Sex		Age at onset	Music type		Seizure type	Duration	Symptoms
86	Ackerman and Banks [43]	66,M	—	—	Unspecific	—	—	—	—
87	Wieser et al. [44]	32,F	—	18	Favourite Italian songs	—	Nocturnal generalized seizures	—	Pleasing female murmuring voices, which took increasing possession of her mind
88	Nakano et al. [45]	23,F	—	19	American pop music	Yes	Yes (unspecified)	3'	—
89	[46]	43,M	—		"The X-files" theme song	No	—	—	—
90	[5]	48,F	No	32	"Arabesque"	Yes	Sleep seizures	4–5'	Epigastric discomfort
91	Gelisse et al. [47]	39,F	Singer	37	Unspecific	No	No	—	Anxiety and tearfulness
92	Lin et al. [48]	6 months,M	—	6 months	Loud music, Beatles in particular	—	Partial seizures	—	—
93	Morocz et al. [49]	48,F	No	42	Whitney Huston and Boyz II men	—	No	—	Pressure in the abdominal and then pectoral area, "rushing" sensation, palpitations, heart racing
94	Wieser et al. [50]	-,F	—	—	Unspecific	—	Yes (unspecified)	—	—
95	Wieser et al. [50]	-,-	—	—	—	—	—	—	—
96	Wieser et al. [50]	-,-	—	—	—	—	—	—	—

(continued)

Table 8.1 (continued)

ID	Authors	Age, sex	Music training	onset (age)	Music genre	Emotional component	Other seizures	Latency	Aura type
97	Wieser et al. [50]	-,-	–	–	–	–	–	–	–
98	Wieser et al. [50]	-,-	–	–	–	–	–	–	–
99	Wieser et al. [50]	-,-	–	–	–	–	–	–	–
100	Wieser et al. [50]	-,-	–	–	–	–	–	–	–
101	Stern J et al. [51]	46,F	–	–	–	No	–	–	–
102	Sacks O [52]	-,M	–	–	–	Yes	–	–	State of intense attention in music listening
103	Sacks O [52]	-,F	–	–	Neapolitan songs	No	–	–	–
104	Anneken [53]	48,F	No	41	Sorrowful lyrics and predominant instrumental background	Yes	Generalized T-C seizure	2'	Epigastric aura, deja vu
105	[54]	49,M	No	32	Favourite music piece	Yes	Yes (unspecified)	–	–
106	[55]	20,F	No	5	Popular music rhythms	–	Yes (unspecified)	–	Internal rhythm perception that quickly evolved into a musical tune followed by a fearful sensation

107	[55]	24,F	Keyboard player	18	Shania Twain's ballad	Yes	No	–	"Tingling, vibrating" feeling in the right side of head quickly followed by an unpleasant abdominal sensation, nausea, lightheadedness.
108	[55]	19,F	No	5	Church hymns	–	Yes (unspecified)	5–10'	Smelling of a not describable odor
109	Claassen DO et al. [56]	65,F	No	46	Slow, melancholic music	Yes	–	–	Fear, tachicardia, crying
110	Cho JW et al. [57]	34,F	No	32	Ballad music at onset, later any type of music	Yes	No	1–5'	Palpitations (no ECG correlate), unpleasant feeling
111	Pittau et al. [58]	36,M	Amateur guitarist	24	Listening or playing music with a strong emotional charge	Yes	No	98–120''	–
112	Mehta et al. [59]	24,F	No	–	Music with a strong rhythmic component	No	Yes (unspecified), preceding musicogenic by 2 years	–	Unpleasant odor, "tingling in the head," and vertigo

(continued)

Table 8.1 (continued)

ID	Authors	Age, sex	Music training	onset (age)	Music genre	Emotional component	Other seizures	Latency	Aura type
113	Marrosu et al. [60]	28,F	No	14	Specific song	No	Yes (unspecified), preceding musicogenic	–	"Funny feelings in the throat"
114	Duanyu et al. [61]	16,M	No	14	Popular rhythms, a song in particual (both listening and singing)	–	Yes (unspecified)	–	Tinnitus, fear
115	Sanchez-Carpintero [62]	7,M	No	5	Music of electronic nature in particular	–	Generalized T-C seizure	5″	–
116	[63]	32,M	No	–	"Russian chanson"	Yes	No	64″	Pleasant feeling, which turns into a non-specific aura with an uncomfortable feeling
117	Seidi [64]	40,M	–	–	Dance traditional music	–	Temporal lobe epilepsy	–	–
118	Seidi [64]	53,M	–	–	Dance traditional music	–	–	–	–
119	Seidi [64]	-,F	–	–	Popular music	–	–	–	–
120	Tezer [65]	33,F	No	17	Affective music in native language	Yes	Yes	2–5′	Boring sensation
121	Wang et al. [66]	42,M	Composer	11	Familiar songs	Yes	Yes (unspecified)	2′	Buzzing/muffled sounds

#	Reference	Age, Sex		Onset	Music		Seizures		Other
122	Klamer et al. [67]	22,M	No	6	Rap music	–	Yes (unspecified)	152'ꞏ	Déjà-vu
123	Cheng [68]	42,M	No	39	Music listened to under previous stressful conditions	Yes	Yes (unspecified), preceding musicogenic	120'ꞏ	Déjà-vu
124	Nagahama Y et al [69]	17,M	Viola player	11	Unspecific	No	Yes (unspecified)	–	–
125	[70]	35,M	No	26	"Moonlight" by Cyndi Wang	–	Sleep seizures	–	–
126	[70]	36,F	Amateur pianist	32	Unspecific	–	Generalized T-C seizure	–	–
127	[70]	55,M	No	28	"Grandma words" by Ricky Hsiao	–	Sleep seizures	–	–
128	Falip et al. [71]	63,M	–	39	Flamenco music in particular	Yes	Yes (unspecified), preceding musicogenic	14'	Epigastric aura
129	Falip et al. [71]	39,M	Amateur guitarist	30	Playing rick music on guitar	Yes	–	–	Epigastric sensation, déjà-vu and fear. Sometimes olfactory sensation
130	Falip et al. [71]	39,F	–	Childhood	Lullabies singed by mother in particular	Yes	Yes (unspecified)	–	Sound distortion

(continued)

Table 8.1 (continued)

ID	Authors	Age, sex	Music training	onset (age)	Music genre	Emotional component	Other seizures	Latency	Aura type
131	Pelliccia et al. [72]	27,F	No	17	Italian pop melodic songs	Yes	Yes (unspecified)	15–28''	Epigastric aura, nausea, tachycardia, déjà-vu
132	[11]	38,F	No	20	Italian pop melodic songs	No	Generalized T-C seizure	–	–
133	[11]	50,F	No	47	Italian and English pop songs	Yes	No	7'	–
134	[73]	19,F	Singer	15	Unspecific songs and singing	–	Yes (unspecified)	–	–
135	[74]	25,M	No	18	Pop	–	Yes (unspecified)	–	–
136	[74]	36,F	No	36	Country	–	Yes (unspecified)	–	–
137	[74]	31,F	No	17	Pop	–	Yes (unspecified)	–	–
138	[74]	28,F	No	23	Unspecific	–	Yes (unspecified)	–	–
139	[74]	61,F	No	46	Melanchonic, church hymns	Yes	Yes (unspecified)	–	–
140	[74]	26,F	No	23	Pop, soft rock	–	Yes (unspecified)	–	–
141	[74]	34,F	No	4	Unfamiliar hymns, classical	–	Yes (unspecified)	–	–
142	[74]	36,F	No	34	Pop and techno	–	Yes (unspecified)	–	–

ID	Reference	Age, Sex	Interictal EEG	Ictal EEG	Age	MRI/fMRI	Type of music	PET	SPECT	MEG	Hypotised EZ
143	[74]	52,F	No	No	14	–	Organ music	–	Yes (unspecified)	–	–
144	Al-Attas et al. [75]	50,F	No	No	36	–	Arabic music in a specific tone	–	–	–	Olfactory
145	Benoit et al. [76]	-;-	–	–	–	–	–	–	Yes (unspecified)	–	–
146	[77]	25,M	Yes	Yes	–	–	Hard rock music	No	Yes (unspecified)	–	Déjà-vu

ID	Seizure type	Interictal EEG	Ictal EEG	MRI/fMRI	PET	SPECT	MEG	Hypotised EZ
1	–	–	–	–	–	–	–	–
2	Generalized	–	–	–	–	–	–	–
3	–	–	–	–	–	–	–	–
4	Simple partial; secondary generalized	–	–	–	–	–	–	–
5	–	–	–	–	–	–	–	–
6	–	–	–	–	–	–	–	–
7	–	–	–	–	–	–	–	–
8	–	–	–	–	–	–	–	–
9	–	–	–	–	–	–	–	–
10	–	–	–	–	–	–	–	–
11	–	–	–	–	–	–	–	–
12	–	–	–	–	–	–	–	–
13	–	–	–	–	–	–	–	Right temporal

(continued)

Table 8.1 (continued)

ID	Seizure type	Interictal EEG	Ictal EEG	MRI/fMRI	PET	SPECT	MEG	Hypotised EZ
14	Generalized	–	–	–	–	–	–	–
15	Generalized	–	–	–	–	–	–	Temporal lobe
16	–	–	–	–	–	–	–	–
17	–	–	–	–	–	–	–	–
18	Generalized	–	–	–	–	–	–	–
19	Secondary generalized	–	–	–	–	–	–	Temporal lobe
20	–	–	–	–	–	–	–	–
21	–	–	–	–	–	–	–	–
22	Secondary generalized	–	–	–	–	–	–	Right temporal
23	Secondary generalized	–	–	–	–	–	–	–
24	–	–	–	–	–	–	–	–
25	–	–	–	–	–	–	–	–
26	–	–	–	–	–	–	–	–
27	–	–	Non focal	–	–	–	–	–
28	–	–	Non focal	–	–	–	–	–
29	–	–	Non focal	–	–	–	–	–
30	–	–	Left parietal	–	–	–	–	Left temporal
31	Generalized	"General dysrhythmia"	Right temporal abnormalities	–	–	–	–	Right temporal
32	–	Right temporo-parietal	–	–	–	–	–	Right temporal
33	Generalized	–	–	–	–	–	–	Left precentral

34	Generalized	"General dysrhythmia"	Left pre- and post-central focus	–	–	–	Left pre/post-central
35	–	–	Right temporal	–	–	–	Right temporal
36	–	–	Right temporal	–	–	–	Right temporal
37	–	Right mediobasal temporal	–	–	–	–	–
38	–	–	Right frontotemporal	–	–	–	Right frontotemporal
39	–	–	–	–	–	–	–
40	–	–	Left temporal	–	–	–	Left temporal
41	–	–	Right temporal	–	–	–	Right temporal
42	–	–	Right frontotemporal	–	–	–	Right frontotemporal
43	–	–	Right temporal	–	–	–	Right temporal
44	–	–	Left temporal	–	–	–	Left temporal
45	–	–	Left temporal	–	–	–	Left temporal
46	Complex partial; generalized	Left temporal slow waves	Generalized dysrhythmia	–	–	–	–
47	Generalized	Normal	Generalized rhythmic discharges	–	–	–	Temporal lobe
48	Complex partial	–	Right temporal	–	–	–	Right temporal
49	–	–	Normal	–	–	–	–

(continued)

Table 8.1 (continued)

ID	Seizure type	Interictal EEG	Ictal EEG	MRI/fMRI	PET	SPECT	MEG	Hypotised EZ
50	–	Normal	–	–	–	–	–	–
51	–	–	–	–	–	–	–	–
52	–	–	Right frontotemporal	–	–	–	–	Right frontotemporal
53	–		Right temporal	–	–	–	–	Right temporal
54	–		Right temporal	–	–	–	–	Right temporal
55	–		Bitemporal	–	–	–	–	–
56	–		–	–	–	–	–	–
57	–		Bitemporal	–	–	–	–	–
58	–		Right temporal	–	–	–	–	Right temporal
59	–	Normal	–	–	–	–	–	–
60	–	Small abnormalities	–	–	–	–	–	–
61	–	Temporal lobe spikes (unknown side)	–	–	–	–	–	–
62	Complex partial	Moderate theta activity in temporal regions	Left temporal	–	–	–	–	Left temporal
63	Complex partial	Recurrent bitemporal delta activity	Bi-temporal sharp activity (R > L)	–	–	–	–	Right temporal
64	–		Bitemporal anterior	–	–	–	–	Left temporal
65	–		–	–	–	–	–	–
66	–	Left fronto-temporal	–	–	–	–	–	–

67	–	–	Left temporal	–	–	–	Left temporal
68	–	–	Left temporal	–	–	–	Left temporal
69	Secondary generalized	Right temporal	Right temporal, with secondary generalization	–	–	–	–
70	–	–	Left temporal	–	–	–	Left temporal
71	–	–	Left temporal	–	–	–	Left temporal
72	–	Right temporo-occipital	–	–	–	–	–
73	Complex partial	–	Right temporal	–	–	–	Right temporal
74	–	–	Left temporo-parietal	–	–	–	Left temporo-parietal
75	–	Bitemporal abnormalities	–	–	–	–	–
76	–	–	Right temporal	–	–	–	Right temporal
77	–	–	Non focal	–	–	–	–
78	Complex partial	Intermittent bitemporal theta activity	Bitemporal theta activity (L > R)	–	–	–	Bilateral temporal
79	Simple partial	–	Right temporo-frontal	–	–	–	–
80	–	Right temporal abnormalities	–	–	–	–	–
81	–	–	Left temporal	–	–	–	Left temporal

(continued)

Table 8.1 (continued)

ID	Seizure type	Interictal EEG	Ictal EEG	MRI/fMRI	PET	SPECT	MEG	Hypotised EZ
82	Generalized	Normal	Bilateral, generalized	–	–	–	–	–
83	Simple partial	Normal	Left temporal	–	–	–	–	Left temporal
84	–	Left temporal abnormalities	–	–	–	–	–	–
85	Simple partial	–	Right mesial temporal	–	–	–	–	–
86	Complex partial	–	–	–	–	–	–	–
87		Fronto-temporal bilateral focal abnormalities	Right frontotemporal	Normal	Right temporal and frontal slight hypometabolism (interictal)	Right temporal hyperperfusion (interictal). Right temporal posterior hypoperfusion, increased right anterior temporopolar hyperperfusion (ictal)	–	Right temporal
88	Complex partial	–	Right temporal (11 Hz rhythmic epileptiform activities) and subsequent delta activities over the right hemisphere	Normal	–	–	–	Right temporal

			(subdural EEG)					
89	–	–	–	Right temporal cortex thickening	–	–	–	Right temporal superior gyrus
90	Complex partial	Bitemporal sharp wave discharges, predominant on the right side	Right temporal (high voltage sharp and slow sharp waves and spikes), then became generalized	Normal	–	Normal (interictal). Right anterior and mesial temporal hyperperfusion (HMPAO, ictal)	–	Right temporal
91	Simple partial	Independent slow waves over the temporal regions	Right temporal	Normal	–	Right temporal hyperperfusion (HMPAO, ictal)	–	Right temporal
92	Simple partial, generalized	Normal	Left temporal	Normal	–	Left temporal hyperperfusion (ictal)	–	Left temporal
93	Complex partial	Normal	Left anterior	Normal (MRI). Right gyrus rectus, ventral frontal lobes, left temporal lobe increased activity (fMRI)	–	Left temporal hypoperfusion (interictal). Left temporal hyperperfusion (ictal)	–	Right gyrus rectus
94	–	–	Right temporal	–	–	–	–	Bilateral temporal
95	–	–	–	–	–	–	–	–
96	–	–	–	–	–	–	–	–
97	–	–	–	–	–	–	–	–
98	–	–	–	–	–	–	–	–

(continued)

Table 8.1 (continued)

ID	Seizure type	Interictal EEG	Ictal EEG	MRI/fMRI	PET	SPECT	MEG	Hypotised EZ
99	–	–	–	–	–	–	–	–
100	–	–	–	–	–	–	–	–
101	–	–	–	Normal	Normal (interictal)	–	Righ temporal	Right temporal
102	Complex partial	–	–	–	–	–	–	Temporal lobe
103	–	–	–	–	–	–	–	–
104	Simple partial	Intermittent theta activity and sharp wave like transients in the left fronto-temporal region	–	Left fronto-temporal astrocitoma	–	Left temporal hyperactivity (IMT, interictal)	–	–
105	Simple partial	Focal spikes in F8 and T4	Normal	Normal	–	Normal (interictal)	–	Right temporal
106	Simple partial	–	Right fronto-temporal activity (4-Hz sharp waves) evolving into high-frequency low-voltage fast activity (IC EEG)	T2 iperintense occipital lesion	Normal (interictal)	–	–	Right anterior lateral temporal

107	Simple partial	–	Right anteromesial temporal (rhythmic intermixed gamma/beta activity, IC EEG)	Normal	Right temporal mild hypometabolism (interictal)	Normal (interictal)	–	Right mesial temporal
108	Complex partial	–	Left mesial temporal; right mesial temporal lobe	Normal	Normal (interictal)	–	Left mesial temporal	Reft and right mesial temporal regions
109	Partial	Right temporal spike discharges	Right fronto-temporal sharp theta and alpha frequency discharges, with rapid spread to the left temporal region.	–	–	–	–	Right fronto-temporal
110	Complex partial	Sharp waves during sleep on the right temporal lobe	Right temporal (rhythmic theta activity)	Normal (MRI) right insula, anterior temporal lobe, amygdala, and hippocampal head BOLD activation (fMRI)	Right insula, amygdala and hippocampal head and anterior temporal hypometabolism (interictal)	Normal (interictal), Right anterior and mesial temporal hyperperfusion (SISCOM, ictal)	–	Right temporal

(continued)

Table 8.1 (continued)

ID	Seizure type	Interictal EEG	Ictal EEG	MRI/fMRI	PET	SPECT	MEG	Hypotised EZ
111	Simple partial	Epileptiform abnormalities over right temporal regions	Right temporal	Normal (MRI). Right fronto-temporo-occipital BOLD activation (fMRI)	–	–	–	Right temporal
112	Complex partial	Bilateral spiking	Right mesial temporal (subdural IC-EEG)	Normal	Right lateral temporal lobe hypometabolism (interictal). Right anteromesial temporal lobe hypermetabolism and right lateral temporal lobe hypometabolism (ictal).	–	–	–
113	Complex partial	–	Right temporal (short sequences of medium voltage theta waves)	Right frontal, right middle-posterior temporal BOLD activation (fMRI)	–	–	–	Right temporal
114	Complex partial, generalized	Left temporal, left frontal	Left frontotemporal	Normal	–	Left temporal and left mesial temporal hypoperfusion (interictal)	–	Left temporal lobe

115	Secondary generalized	—	Bitemporal sharp waves spreading to centrotemporal areas	—	—	—	—	—
116	Complex partial	Left temporal spikes	Left temporal	Normal (MRI). "Snowballing" bilateral BOLD activation (fMRI)	—	—	—	Left temporal
117	—	—	—	—	—	—	—	—
118	Generalized	Normal	Normal	Normal	—	—	—	—
119	Secondary generalized	—	Left temporal	—	—	—	—	Left temporal
120	Complex partial	—	Right ippocampus (musicogenic), left ippocampus (spontaneous seizure). Intracranial EEG	Mild atrophy of the left anterior temporal lobe	Right temporal hypometabolism (interictal)	—	—	Right hippocampus (intracranial EEG) for musicogenic seizure, left hippocampus (intracranial EEG) for spontaneous seizure

(continued)

Table 8.1 (continued)

ID	Seizure type	Interictal EEG	Ictal EEG	MRI/fMRI	PET	SPECT	MEG	Hypotised EZ
121	Complex partial	Intermittent left temporal slowing waves	Left temporal (intracranial EEG)	Normal	Left temporal hypometabolism (interictal)	Left lateral superior temporal gyrus and contiguous supratemporal plane hyperperfusion (ictal)	Left temporal	Left temporal
122	Secondary generalized	–	Right hippocampus with fast propagation to the mesial frontal depth electrode (intracranial EEG)	Normal (MRI). Right mesial temporal lobe (right hippocampus extending to the right insula and the right amygdala) BOLD activation (fMRI)	–	–	Right mesial frontal, bilateral mesial temporal	Right mesial temporal
123	Secondary generalized	Left frontotemporal sharp waves	Left frontotemporal	Normal	–	–	–	Left fronto-temporal
124	Partial	–	Right temporal	Normal	Normal (ictal)	–	Right fronto-temporo-insular	Right temporal
125	Complex partial	Normal	Left mesial temporal	Normal	–	–	–	Left mesial temporal

126	Simple partial	Normal	Left mesial temporal, neocortical temporal	Normal	–	–	–	Left mesial temporal
127	Complex partial	Fronto-temporal discharges	Left mesial temporal	Right anterior temporal encephalomalacia	–	–	–	Left mesial temporal
128	Complex partial	Normal	Right temporal	Normal	Right medial temporal hypometabolism (interictal)	–	–	Right temporal
129	Simple partial; complex partial	Left temporal discharges	–	Normal	Bilateral temporal and insular hypometabolism (interictal)	–	–	–
130	Complex partial	–	–	Normal	Right medial temporal hypometabolism (interictal)	–	–	Right temporal
131	Complex partial	Right temporal spikes	Right temporal and central (T4, C4). SEEG:Right amygdala and hippocampus	–	–	–	–	Right temporal
132	Complex partial	Left fronto-temporal spikes	–	Normal	–	–	–	Left fronto-temporal
133	Complex partial	Right fronto-temporal spikes	Right fronto-temporal	Normal	–	–	–	Right fronto-temporal

(continued)

Table 8.1 (continued)

ID	Seizure type	Interictal EEG	Ictal EEG	MRI/fMRI	PET	SPECT	MEG	Hypotised EZ
134	Simple partial	Normal	Normal	Right temporal lesion	–	–	–	Right temporal
135	Complex partial	Bitemporal abnormalities	Right temporal	Normal	–	–	–	Right temporal
136	Simple partial	Bitemporal abnormalities	–	Normal	–	–	–	Left temporal
137	Complex partial	Right temporal	–	Normal	–	–	–	Right temporal
138	Simple partial	Right temporal	–	Normal	–	–	–	Right temporal
139	Complex partial	Bitemporal abnormalities	Right temporal	Normal	–	–	–	Right temporal
140	Complex partial	Normal	Left temporal	Normal	–	–	–	Left temporal
141	Complex partial	Bitemporal abnormalities	–	Normal	–	–	–	Right temporal
142	Complex partial	Normal	Right temporal	Normal	–	–	–	Right temporal
143	Complex partial	Bitemporal abnormalities	–	Right mesial temporal sclerosis	–	–	–	Right temporal
144	Complex partial; generalized	–	Left temporal	Normal	–	–	–	Left temporal
145	–	–	Right temporal	–	–	–	–	Right temporal
146	Simple partial	–	Right temporal	Bilateral tempo-mesial abnormalities	–	–	–	Bilateral temporal

EEG electroencephalogram; *EZ* epileptogenic zone; *fMRI* functional magnetic resonance imaging; *IC-EEG* intracranial EEG; *MEG* magnetoencephalography; *MRI* magnetic resonance imaging; *PET* positron emission tomography; *SPECT* single-photon emission computed tomography

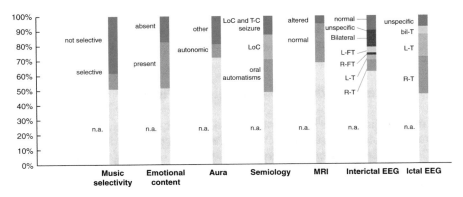

Fig. 8.1 Graphical representation of the reviewed cases of musicogenic epilepsy. Bars include percentage representation of the following findings: selectivity of music in provoking seizures, emotional charge of epileptogenic stimulus, aura's features, seizure's semiology, structural magnetic resonance imaging (MRI) features, and interictal and ictal electroencephalographic (EEG) findings (*n.a.* not available; *LoC* loss of consciousness; *T-C* tonic clonic; *L-T* left temporal; *R-T* right temporal; *L-FT* left frontotemporal; *R-FT* right frontotemporal; *bil-T* bilateral temporal)

8.2   Clinical Features of People with Musicolepsia

The demographical picture of people suffering from musicolepsia displays a mean age at observation (reported in 125 cases) of 38 ± 14 years, with mean seizure onset (reported in 69 cases) at 26 ± 13 years, without significant unbalancing in terms of gender (specified in 137 cases, 55% F, 45% M). Musical stimuli are not selective as seizures trigger in most cases (*n* = 87). In the remaining observations, only 16 cases (27%) reported seizures provoked exclusively by music listening. Other kinds of seizures coexisting with musicolepsia span from spontaneous seizures (*n* = 23, 39%) to non-musicogenic reflex seizures (*n* = 4, 3%) and other not detailed seizures (*n* = 17, 29%).

Since extensive musical training is known to induce long-lasting changes in cortical excitability [78], it would be speculated that music expertise may affect the probability of undergoing musicogenic seizures. However, the low prevalence of trained musicians among people with musicolepsia (only 13 cases, 9% of total) makes this hypothesis unlikely.

Feelings prodromal to the musicogenic seizures (i.e., aura, reported in 41 patients) is often autonomic (14 patients), but other symptoms ranging from auditory, olfactory, visual to somatosensory phenomena are also described. In some cases, an intense sense of fear starts at music onset and precedes the seizure: this would be the literary case reported in the manuscript *"fear of music"* by the Russian writer B. Nikonov and subsequently reported by Bekhterev [79].

Supporting the ictogenic involvement of temporal lobe, seizure semiology (reported in 83 cases) displays oroalimentary automatism (chewing, swallowing, smacking lips) in about one-fourth of observations (23 patients). Loss of consciousness represents a common event, being reported in 49 cases (59%), 16 of them (19%) including tonic-clonic seizures.

In most cases, musicolepsia is not secondary to morphological brain alterations. In fact, when performed (46 cases), structural MRI is usually unremarkable (90%).

The rare structural abnormalities usually involve the temporal lobes (7 cases; [46, 53, 65, 70, 73, 74, 77]), in a single case extending to frontal areas [53]. Finally, in one patient, unspecific occipital abnormalities were described [55].

8.3   Is the *Emotional Glow* to Light the Fire?

One of the main reasons why humans are looking for musical experience is for its natural vocation to arouse emotions: the obsessive rhythms of tribal rituals provoke emotional bursts and *trance* experiences [80]; ragas belonging to Indian classical music are prescribed to evoke a precise emotional effect [81]; the administration of music linked to specific memories can resurface their affective burden (see the effects of Sam's playing *"As time goes by"* in the movie *Casablanca*, 1942).

Despite its key role in musical experience, the presence/absence of emotional content of musical stimulus triggering seizures has not been methodically investigated. Among the 71 cases where it has been enquired, 46 (65%) reported a salient emotional charge related to the musical stimulus. Of note, the presence of such an affective component is not clearly inferable in about half of cases, leading to a potential underestimation of the presence of emotional factors. Besides incomplete case-history collections, also patients' difficulties in remembering seizure events [13] as well as reluctance to expose personal anamnestic details [82] may have sometimes hindered the access to this kind of information.

But, how to determine the association between such an emotional component and seizure ignition?

A first way could be investigating if the epileptogenic zones associated with "emotionally charged" seizures are more frequently situated in the right hemisphere, this latter traditionally acknowledged as "dominant" for emotional processing.

Selecting data of cases in which both emotional content and epileptogenic zone are defined ($n = 41$, 3 bilateral Ictal EEG excluded), patients with an emotional component ($n = 30$) showed a higher prevalence of right-sided epileptic zone (22 vs 7). However, a chi-square test performed to detect if epileptogenic-zone laterality is biased by the presence/absence of such an emotional component did not return any significant result (chi = 0.016, $p > 0.05$); in other words, also people whose seizures are elicited by *"cold"* music showed a predominance of right-hemisphere epileptogenic zones.

These findings are consistent with the view that the processing of emotional content of music, as well as the mapping of basic—potentially ictogenic—sound features (e.g., pitch, timbre, and melodic contour) are prerogative (although not exclusive) functions of the right hemisphere. This view is supported by a huge amount of behavioral studies [83, 84], neuroimaging investigations [85, 86], and clinical data from patients with brain lesions [87, 88] and music-related disturbances [89].

Music, however, it's not a mere matter of pitch, rhythm, and—when present— "emotional glow". Indeed, in order to be noticed as *musical*, sequences of discrete elements need to be endowed with a hierarchical structure [90], i.e., with a *syntax* [91]. This prerequisite extends to multiple musical cultures: from the complex rhythmic organization of traditional sub-Saharan music to the choice of the proper *svara* within Indian *ragas*, as well as in the *armonic* structure of western musical pieces. As in spoken language, neural processing of music syntax predominantly involves the left hemisphere [92–94], which is strongly recruited during the perception of sounds associated with spoken words [95].

Investigations on patients with musicolepsia can indirectly support such hemispherical bias: Tseng et al. [70] administered different versions of epileptogenic songs (i.e., with/without lyrics) to three patients suffering from left-temporal musicogenic epilepsy, demonstrating that only vocal versions of songs were able to provoke seizures. These experimental findings are consistent with the previous musicolepsia literature: music containing lyrics provokes seizures in the left-temporal epileptogenic zone twofold more frequently than the right temporal region, thus indicating that language-associated elements may represent a key triggering factor for left-temporal musicogenic seizures.

However, the generalization of such an observation needs to be considered with caution: even music with lyrics may conceal epileptogenic triggers within non-linguistic elements, spanning from rhythmic riffs to instrumental solos. Exemplary in this regard is the case described by Bekhterev [79] about the Russian writer B. Nikonov, who experienced a seizure at the theatre while listening "The Prophet" by Meyerbeer. Although this opera is composed by wide lyrical segments along all its five acts, the seizure—preceded by an intense fear feeling—just occurred during the listening of an instrumental part (i.e., the ballet introducing the third act).

Another element that supports the presence of multiple stages of high-order processes during music listening is the long latency between the epileptogenic stimulus and seizure onset. This aspect represents a typical feature distinguishing musicogenic seizures from other reflex seizures [3, 11]. Noteworthy, patients with an epileptogenic zone in the right hemisphere show higher latencies from stimulus onset (279 ± 236 s vs 82.2 ± 43.3 s, Mann-Whitney $U = 6$; $p = 0.023$), possibly reflecting the functional predominancy of the right hemisphere in the emotional processing of music [10].

Even if it is informative about the "emotional" and "linguistic" domains of music, the mere hemispheric laterality of the epileptogenic zone is not sufficient to detail the neural substrates of music processing. For this purpose, the integration of neurophysiological (e.g., electroencephalography—EEG) and functional neuroimaging (fMRI, PET, SPECT) data give the opportunity to associate clinical symptoms of people with musicolepsia with brain topographies, thus providing insights into the functional role of brain regions involved in music processing.

8.4   Neurophysiological and Functional-Neuroimaging Findings of Musicolepsia

The introduction in the first half of twentieth century of the clinical electroencephalography (EEG) represented a milestone for epileptology. In addition to providing temporal and spatial information about the brain substrates of seizures, the possibility to demonstrate neurophysiological abnormalities during fits gave the opportunity to enclose in the neurological realm paroxysmal events (e.g., musicolepsia), which until then were often classified as "hysterical" manifestations [13].

From the first EEG recording of a seizure evoked by mean of a gramophone [13], 84 ictal EEG have been performed to date in people with musicolepsia, of which 40 (48%) displayed right-temporal alterations, 24 (29%) left temporal, 7 (8%) bitemporal, and 11 (13%) showed not focal or unspecific alterations. Intracranial EEG, performed on 10 subjects, found an epileptogenic zone in the temporal lobe (8 right, 1 left, 1 bilateral), involving temporo-mesial structures in 6 cases. Moving to neuroimaging techniques, the ictal SPECT study (7 cases) displayed 4 right temporal (of which two included mesial structures) and 3 left temporal alterations. Ictal fMRI (6 cases) showed right-sided increased activation in 5 cases, of which 2 limited to temporal regions and 3 involving also frontal or insular regions. Finally, one case [63] showed bilateral activation in a "snowballing-like" pattern (see Fig. 8.1). Seizure latency from music onset was indicated in 18 cases (210 ± 210 s).

The long latency between musical stimulus and seizure onset makes musicogenic seizures as a use case to investigate the brain mechanisms responsible for the ignition of the ictal phases. In this regard, the study of interictal EEG findings may help identify epileptic focus. Moreover, additional information about their morphology, rate, time delay, spread etc. may help to differentiate areas of origin from areas of propagation of interictal discharges [96].

Interictal EEG abnormalities have been reported in 55 cases of musicolepsia (see Fig. 8.1). Among them, 13 showed right-temporal, 11 left-temporal, 13 bi-temporal, and 1 case reported "temporal spikes" without side specification. The detection of neurophysiological alterations was extended to frontal electrodes in 9 cases (6 left, 2 right, 1 bilateral). Finally, 3 cases displayed unspecific EEG alterations, and 11 recordings were unremarkable. The morphology of interictal alterations (reported in 15 cases) included sharp waves ($n = 5$), slow waves ($n = 2$), and spikes ($n = 8$). In one case [54], the detailed study of interictal spikes resulted in determinants to define the epileptogenic zone since no ictal data were available.

The relationship between interictal spiking and epileptogenesis is still a matter of debate [96]. On the one hand, it has been demonstrated that an increased cortical excitability favors epileptic spikes, whose spatiotemporal density may reach a critical threshold able to provoke seizures [97–99], so that there would be a direct association between the sustained presence of interictal abnormalities and "epileptogenity" [100]. On the other side, several studies demonstrated a reduction in the interictal discharge rate before seizure onset, suggesting that—at least in some cases—spikes may exert a "protective" role toward seizure onset [96, 101, 102]. This view is supported by animal experimental models in which interictal activity is able to reduce the ability of the entorhinal cortex to generate ictal events [103], as well as by pharmacological investigations that evidenced an opposite dynamic between interictal and ictal spike rates following the administration of $GABA_B$ agonists [104].

These findings are consistent with a recent high-density EEG report of a patient with musicolepsia, in which seizures were evoked by listening to epileptogenic music [11]. Here, independent component analysis (ICA) was applied to identify epileptiform markers and to detect putative epileptogenic sources: Ictal and

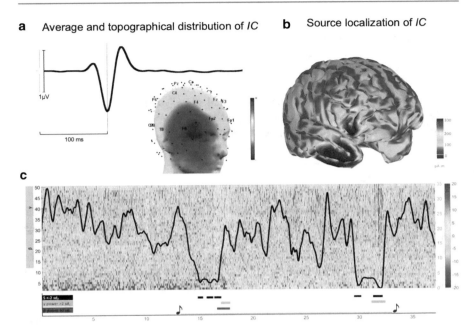

a Average and topographical distribution of *IC*   **b** Source localization of *IC*

Fig. 8.2 Neurophysiological features of interictal spikes and temporal dynamic of their density during EEG recording, patient R.C., adapted from Nuara et al., Clinical Neurophysiology, 2020, 131 (10), 2393–2401 with permission **Panel A:** Average spike of the independent component and its topography. **Panel B**: Source localization of IC is displayed onto a cortical sheet template. Note the right frontotemporal localization. **Panel C**: Temporal dynamic (minutes) of spike density (black line) overlapped to time-frequency plot of EEG recording. The first spike-density negative peaks (indicated by the first three black rectangles on the bottom panel) just precede seizure onset, this latter associated with the expected increase of beta and gamma power (respectively, red and green rectangles). The second spike-density negative peak precedes an increase in gamma power, without overt clinical seizure correlates. The musical notes in the timeline indicate the start and the end of the music-listening period

interictal discharges highlighted a right frontotemporal localization and a suppression of spike density preceding the seizure onset (see Fig. 8.2).

The involvement of extra-auditory domains as potential sources in musicogenic seizures is consistent with the complex architecture of neural substrates of music listening. Indeed, from the primary auditory cortices involved in tonal mapping, distinct—but interacting—pathways spread along two main streams: the dorsal-posterior one interfaces with the inferior parietal cortex, while the ventral-anterior goes toward the anterior part of the temporal lobe [105]. Both streams interact with distinct frontal areas, forming frontotemporal cortical loops that play a key role in maintaining musical information in working memory, controlling musical expectancy [7].

When, during music listening, we experience an emotional peak (i.e., the *"music chills"*), bursts in autonomic activity (e.g., increase of heart rate) are associated with an increased activity in anterior frontal regions, ventral striatum, mesolimbic regions, and other key subcortical structures involved in reward and emotion [10]. Moreover, the functional connectivity between auditory and frontal areas increases as a function of increasing pleasure component during music listening ([106], see also for a review [7]).

These findings suggest that the "emotional glow" of perceived music—especially when reported by patients—may represent an independent trigger for musicogenic seizures, possibly involving extra-auditory regions.

8.5 Conclusions

Eighty-five years have passed since Critchley first adopted the term "musicolepsia" to describe seizures induced by music listening. Since then, the growing knowledge in the field of neuroscience and the technological advances in neuroimaging allowed us to rethink musicogenic seizures from a simple *"operation of a conditioned reflex in a Pavlovian sense"* [1], to a complex phenomenon involving multiple aspects of music processing. Among them, the "emotional glow" of music is endowed with a peculiar faculty to ignite seizures.

Indeed, more than a nosological class in the wide realm of epilepsy, musicolepsia represents a unique window on the neural substrates of human music experience.

References

1. Critchley M. Musicogenic epilepsy. Brain. 1937;60(1):13–27. https://doi.org/10.1093/brain/60.1.13.
2. Critchley M, Henson RA. Music and the brain: studies in the neurology of music. Berlin: Elsevier Science; 2014. http://qut.eblib.com.au/patron/FullRecord.aspx?p=1655618
3. Avanzini G. Musicogenic seizures. Ann N Y Acad Sci. 2003;999:95–102.
4. Fujinawa A, Kawai I, Ohashi H, Kimura S. A case of musicogenic epilepsy. Folia Psychiatr Neurol Jpn. 1977;31(3):463–72.
5. Genç BO, Genç E, Taştekin G, Iihan N. Musicogenic epilepsy with ictal single photon emission computed tomography (SPECT): could these cases contribute to our knowledge of music processing? Eur J Neurol. 2001;8(2):191–4.
6. Poskanzer DC, Brown AE, Miller H. Musicogenic epilepsy caused only by a discrete frequency band of church bells. Brain. 1962;85:77–92.
7. Zatorre RJ, Salimpoor VN. From perception to pleasure: music and its neural substrates. Proc Natl Acad Sci U S A. 2013;110(Suppl 2):10430–7. https://doi.org/10.1073/pnas.1301228110.
8. Kunert R, Willems RM, Casasanto D, Patel AD, Hagoort P. Music and language syntax interact in Broca's area: an fMRI study. PLoS One. 2015;10(11):e0141069. https://doi.org/10.1371/journal.pone.0141069.
9. Gordon CL, Cobb PR, Balasubramaniam R. Recruitment of the motor system during music listening: an ALE meta-analysis of fMRI data. PLoS One. 2018;13(11):e0207213. https://doi.org/10.1371/journal.pone.0207213.
10. Blood AJ, Zatorre RJ. Intensely pleasurable responses to music correlate with activity in brain regions implicated in reward and emotion. Proc Natl Acad Sci. 2001;98(20):11818–23. https://doi.org/10.1073/pnas.191355898.
11. Nuara A, Mirandola L, Fabbri-Destro M, Giovannini G, Vecchiato G, Vaudano AE, Tassinari CA, Avanzini P, Meletti S. Spatio-temporal dynamics of interictal activity in musicogenic

epilepsy: two case reports and a systematic review of the literature. Clin Neurophysiol. 2020;131(10):2393–401. https://doi.org/10.1016/j.clinph.2020.06.028.

12. Vanelle JM. Contribution a l'etude de l'epilepsie musicogenique. These, Paris, University. 1982.

13. Shaw D, Hill D. A case of musicogenic epilepsy. J Neurol Neurosurg Psychiatry. 1947;10(3):107–17. https://doi.org/10.1136/jnnp.10.3.107.

14. Joynt RJ, Green D, Green R. Musicogenic epilepsy. JAMA. 1962 Feb 17;179(7):501–4.

15. Stubbe-Teglbjaerg H. On musciogenic epilepsy. Acta Psychiatrica Scandinavica. 1949 Dec;24(3-4):679–90.

16. Arellano AP, Schwab RS, Casby JU. Sonic activation. Electroencephalogr Clin Neurophysiol. 1950;2(1-4):215–7.

17. Dow RS. Electroencephalographic findings in a case of musicogenic epilepsy. In: Electroencephalogr Clin Neurophysiol 1951 (Vol. 3, No. 3, pp. 384-384).

18. Vercelletto P. A propos d'un cas d'épilepsie musicogénique. Présentation d'une crise temporale, discussion sur son point de départ. Rev Neurol. 1953;88:379–82.

19. Hoheisel HP, Walch R. Ein Fall von akustischer Reflexepilepsie. Psychiatr Neurol Med Psychol. 1953 May;1:194–200.

20. Levi PG. Su un caso di epilessia musicogena. Neuropsichiatria. 1954;10:119–28.

21. Bickford RG. Musicogenic epilepsy. Electroencephalogr Clin Neurophysiol. 1956;8(1):152–3.

22. Weber R. Musikogene epilepsie. Nervenarzt. 1956;27:337–40.

23. Daly DD, Barry MJ Jr. Musicogenic epilepsy: report of three cases. Psychosom Med. 1957 Sep 1;19(5):399–408.

24. Reifenberg E. Beitrag zur Kasuistik der musikogenen Epilepsie. Psychiatr Neurol Med Psychol. 1958 Mar;1:88–91.

25. Barrios Del Risco P, Esslen E. Epilepsia musicógena. Acta Neurol Latinoamer. 1958;4:130–44.

26. Titeca J. Lepilepsie musicogenique: Revue generale a propos d'un cas personnel suivi pendant quatorze ans. Acta Neurol Belg. 1965;65:598–648.

27. Piotrowski A. A case of musicogenic epilepsy-clinical and EEG study. Electroencephalogr Clin Neurophysiol 1959 (Vol. 11, No. 1, pp. 176-176).

28. Bash KW, Bash-Liechti J. Die Psychotherapie eines Falles von musikogener Epilepsie. Schweiz Arch Neurol Psychiatr. 1959;83:196–221.

29. Yvonneau M, de Barros-Ferreira M. A propos de l'épilepsie audiogénique. Presse Med. 1963;71–616.

30. Gornik VM. Apropos of musicogenic epilepsy. Zh Nevropatol Psikhiatr Im S S Korsakova. 1952;1964(64):1227–31.

31. Perpiniotis D. A propos d'un cas d'épilepsie musicogénique pure, influencé par la méthylphényléthylhydantoıne. Rev Neurol. 1964;110:226–31.

32. Forster FM, Klove H, Peterson WG, Bengzon AR. Modification of musicogenic epilepsy extinction technique. Trans Am Neurol Assoc. 1965;90:179–82.

33. Toivakka E, Lehtinen LO. Musicogenic epilepsy: a case report.

34. Dearman HB, Smith BM. A case of musicogenic epilepsy. JAMA. 1965 Sep 27;193(13):1123–5.

35. Gastaut H, Tassinari CA. Triggering mechanisms in epilepsy the electroclinical point of view. Epilepsia. 1966 Jun;7(2):85–138.

36. Vizioli R, Simone F, Felici F. L'epilessia musicogena: contributo clinico. Riv Neurol. 1974;44:448–70.

37. Newman P, Saunders M. A unique case of musicogenic epilepsy. Arch Neurol. 1980 Apr 1;37(4):244–5.

38. Sutherling WW, Hershman LM, Miller JQ, Lee SI. Seizures induced by playing music. Neurology. 1980 Sep 1;30(9):1001–4.

39. Brien SE, Murray TJ. Musicogenic epilepsy. Can Med Assoc J. 1984 Nov 11;131(10):1255.

40. Byun YJ, Hah JS, Park CS. A case of musicogenic epilepsy. J Korean Neurol Assoc. 1989:123–30.

41. Jallon P, Heraut LA, Vanelle JM, Beaumanoir A, Gastaut H, Naquet R, editors. Reflex seizures and reflex epilepsies. Geneva: Éditions Médecine & Hygiène; 1989. p. 269–74.

42. Smeijsters H, van Den Berk P. Music therapy with a client suffering from musicogenic epilepsy: a naturalistic qualitative single-case research. Arts Psychother. 1995;
43. Ackerman RJ, Banks ME. A neuropsychological case study of musicogenic epilepsy. Arch Clin Neuropsychol. 1995;4(10):286–7.
44. Wieser HG, Hungerböhler H, Siegel AM, Buck A. Musicogenic epilepsy: review of the literature and case report with ictal single photon emission computed tomography. Epilepsia. 1997 Feb;38(2):200–7.
45. Nakano M, Takase Y, Tatsumi C. A case of musicogenic epilepsy induced by listening to an American pop music. Rinsho Shinkeigaku. 1998 Dec 1;38(12):1067–9.
46. Trevathan E, Gewirtz RJ, Cibula JE, Schmitt FA. Musicogenic seizures of right superior temporal gyrus origin precipitated by the theme song from 'The X-files'. Epilepsia. 1999;40:23.
47. Gelisse P, Thomas P, Padovani R, Hassan-Sebbag N, Pasquier J, Genton P. Ictal SPECT in a case of pure musicogenic epilepsy. Epileptic Disord. 2003 Sep 1;5(3):133–7.
48. Lin KL, Wang HS, Kao PF. A young infant with musicogenic epilepsy. Pediatr Neurol. 2003 May 1;28(5):379–81.
49. Mórocz IÁ, Karni A, Haut S, Lantos G, Liu G. fMRI of triggerable aurae in musicogenic epilepsy. Neurology. 2003 Feb 25;60(4):705–9.
50. Wieser HG. Musicogenic seizures and findings on the anatomy of musical perception. Reflex epilepsies: progress in understanding. 2004 Apr;13:79–91.
51. Stern JM, Tripathi M, Akhtari M, Korb A, Engel J, Cohen MS. Musicogenic seizure localization with simultaneous EEG and functional MRI (SEM). Neurology 2006 Mar 14 (Vol. 66, No. 5, pp. A90-A90).
52. Sacks O. The power of music. Brain. 2006 Oct 1;129(10):2528–32.
53. Anneken K, Fischera M, Kolska S, Evers S. An unusual case of musicogenic epilepsy in a patient with a left fronto-temporal tumour. J Neurol. 2006;253(11):1502–4. https://doi.org/10.1007/s00415-006-0257-1.
54. Shibata N, Kubota F, Kikuchi S. The origin of the focal spike in musicogenic epilepsy. Epileptic Disord. 2006;8(2):131–5.
55. Tayah TF, Abou-Khalil B, Gilliam FG, Knowlton RC, Wushensky CA, Gallagher MJ. Musicogenic seizures can Arise from multiple temporal lobe foci: intracranial EEG analyses of three patients. Epilepsia. 2006;47(8):1402–6. https://doi.org/10.1111/j.1528-1167.2006.00609.x.
56. Claassen DO, Walting PJ, Tan KM, Pittock PJ, So EL. Elvis and epilepsy; A case of musicogenic epilepsy treated with music. Epilepsia 2007 Oct 1 (Vol. 48, pp. 22-22).
57. Cho JW, Seo DW, Joo EY, Tae WS, Lee J, Hong SB. Neural correlates of musicogenic epilepsy: SISCOM and FDG-PET. Epilepsy Res. 2007 Dec;77(2-3):169–73.
58. Pittau F, Tinuper P, Bisulli F, Naldi I, Cortelli P, Bisulli A, Stipa C, Cevolani D, Agati R, Leonardi M, Baruzzi A. Videopolygraphic and functional MRI study of musicogenic epilepsy. A case report and literature review. Epilepsy Behav. 2008 Nov;13(4):685–92.
59. Mehta AD, Ettinger AB, Perrine K, Dhawan V, Patil A, Jain SK, Klein G, Schneider SJ, Eidelberg D. Seizure propagation in a patient with musicogenic epilepsy. Epilepsy Behav. 2009 Feb;14(2):421–4.
60. Marrosu F, Barberini L, Puligheddu M, Bortolato M, Mascia M, Tuveri A, Muroni A, Mallarini G, Avanzini G. Combined EEG/fMRI recording in musicogenic epilepsy. Epilepsy Res. 2009 Mar;84(1):77–81.
61. Duanyu N, Yongjie L, Guojun Z, Lixin C, Liang Q. Surgical treatment for musicogenic epilepsy. J Clin Neurosci. 2010 Jan;17(1):127–9.
62. Sanchez-Carpintero R, Patiño-Garcia A, Urrestarazu E. Musicogenic seizures in Dravet syndrome. Dev Med Child Neurol. 2013 Jul;55(7):668–70.
63. Diekmann V, Hoppner AC. Cortical network dysfunction in musicogenic epilepsy reflecting the role of snowballing emotional processes in seizure generation: an fMRI-EEG study. Epileptic Disord. 2014;1:31–44. https://doi.org/10.1684/epd.2014.0636.
64. Seidi O, El Sadig S, Ahmed A, et al. J Neurol Sci. 2015;357(Supplement 1):e32.

65. Tezer FI, Bilginer B, Oguz KK, Saygi S. Musicogenic and spontaneous seizures: EEG analyses with hippocampal depth electrodes. Epileptic Disord. 2014;16(4):500–5. https://doi.org/10.1684/epd.2014.0706.
66. Wang ZI, Jin K, Kakisaka Y, Burgess RC, Gonzalez-Martinez JA, Wang S, Ito S, Mosher JC, Hantus S, Alexopoulos AV. Interconnections in superior temporal cortex revealed by musicogenic seizure propagation. J Neurol. 2012 Oct;259(10):2251–4.
67. Klamer S, Rona S, Elshahabi A, Lerche H, Braun C, Honegger J, Erb M, Focke NK. Multimodal effective connectivity analysis reveals seizure focus and propagation in musicogenic epilepsy. Neuroimage. 2015 Jun;113:70–7.
68. Cheng JY. Musicogenic epilepsy and treatment of affective disorders: case report and review of pathogenesis. Cogn Behav Neurol. 2016 Dec;29(4):212–6.
69. Nagahama Y, Kovach CK, Ciliberto M, Joshi C, Rhone AE, Vesole A, Gander PE, Nourski KV, Oya H, Howard MA, Kawasaki H, Dlouhy BJ. Localization of musicogenic epilepsy to Heschl's gyrus and superior temporal plane: case report. J Neurosurg. 2018 Jul;129(1):157–64.
70. Tseng WEJ, Lim SN, Chen LA, Jou SB, Hsieh HY, Cheng MY, Chang CW, Li HT, Chiang HI, Wu T. Correlation of vocals and lyrics with left temporal musicogenic epilepsy. Ann N Y Acad Sci. 2018; https://doi.org/10.1111/nyas.13594.
71. Falip M, Rodriguez-Bel L, Castañer S, Miro J, Jaraba S, Mora J, Bas J, Carreño M. Musicogenic reflex seizures in epilepsy with glutamic acid decarbocylase antibodies. Acta Neurol Scand. 2018 Feb;137(2):272–6.
72. Pelliccia V, Villani F, Gozzo F, Gnatkovsky V, Cardinale F, Tassi L. Musicogenic epilepsy: a stereo-electroencephalography study. Cortex. 2019 Nov;120:582–7.
73. Bass DI, Shurtleff H, Warner M, Knott D, Poliakov A, Friedman S, Collins MJ, Lopez J, Lockrow JP, Novotny EJ, Ojemann JG, Hauptman JS. Awake mapping of the auditory cortex during tumor resection in an aspiring musical performer: A case report. Pediatr Neurosurg. 2020;55(6):351–8. https://doi.org/10.1159/000509328.
74. Smith KM, Zalewski NL, Budhram A, Britton JW, So E, Cascino GD, Ritaccio AL, McKeon A, Pittock SJ, Dubey D. Musicogenic epilepsy: expanding the spectrum of glutamic acid decarboxylase 65 neurological autoimmunity. Epilepsia. 2021;62(5) https://doi.org/10.1111/epi.16888.
75. Al-Attas AA, Al Anazi RF, Swailem SK. Musicogenic reflex seizure with positive antiglutamic decarboxylase antibody: a case report. Epilepsia Open. 2021;6(3):607–10.
76. Benoit J, Martin F, Thomas P. Musicogenic epilepsy with ictal asystole: a video-EEG case report. Epileptic Disord. 2021;23(4):649–54.
77. Morano A, Orlando B, Fanella M, Irelli EC, Colonnese C, Quarato P, Giallonardo AT, Di Bonaventura C. Musicogenic epilepsy in paraneoplastic limbic encephalitis: a video-EEG case report. Epileptic Disord. 2021;23(5):754–9. https://doi.org/10.1684/epd.2021.1322.
78. Schlaug G. Musicians and music making as a model for the study of brain plasticity. Prog Brain Res. 2015;217:37–55. https://doi.org/10.1016/bs.pbr.2014.11.020.
79. Bekhterev V. O reflektornoi epilepsi pod oliyaniem evyookovic razdrazheniye;1914. p. 513.
80. Rouget G. Music and trance: a theory of the relations between music and possession. Chicago: University of Chicago Press; 1985.
81. Valla JM, Alappatt JA, Mathur A, Singh NC. Music and emotion—a case for north Indian classical music. Front Psychol. 2017;8:2115. https://doi.org/10.3389/fpsyg.2017.02115.
82. Gastaut H, Tassinari CA. Triggering mechanisms in epilepsy the electroclinical point of view. Epilepsia. 2010;7(2):85–138. https://doi.org/10.1111/j.1528-1167.1966.tb06262.x.
83. Kallman HJ, Corballis MC. Ear asymmetry in reaction time to musical sounds. Percept Psychophys. 1975;17(4):368–70. https://doi.org/10.3758/BF03199348.
84. Taub JM, Tanguay PE, Doubleday CN, Clarkson D, Remington R. Hemisphere and ear asymmetry in the auditory evoked response to musical chord stimuli. Physiol Psychol. 1976;4(1):11–7. https://doi.org/10.3758/BF03326537.
85. Gaab N, Gaser C, Zaehle T, Jancke L, Schlaug G. Functional anatomy of pitch memory—an fMRI study with sparse temporal sampling. NeuroImage. 2003;19(4):1417–26. https://doi.org/10.1016/S1053-8119(03)00224-6.

86. Zatorre R, Evans A, Meyer E. Neural mechanisms underlying melodic perception and memory for pitch. J Neurosci. 1994;14(4):1908–19. https://doi.org/10.1523/JNEUROSCI.14-04-01908.1994.
87. Berlin CI, Lowe-Bell SS, Jannetta PJ, Kline DG. Central auditory deficits after temporal lobectomy. Arch Otolaryngol Head Neck Surg. 1972;96(1):4–10. https://doi.org/10.1001/archotol.1972.00770090042003.
88. Gordon MC. Reception and retention factors in tone duration discriminations by brain-damaged and control patients. Cortex. 1967;3(2):233–49. https://doi.org/10.1016/S0010-9452(67)80014-5.
89. Stewart L, von Kriegstein K, Warren JD, Griffiths TD. Music and the brain: disorders of musical listening. Brain. 2006;129(10):2533–53. https://doi.org/10.1093/brain/awl171.
90. Schoenberg A. Theory of harmony. New York: Philosophical Library; 1948.
91. Lerdahl F, Jackendoff R. A generative theory of tonal music. Cambridge: MIT Press; 1983.
92. Koelsch S, Gunter TC, Cramon D Yv, Zysset S, Lohmann G, Friederici AD. Bach speaks: a cortical 'language-network' serves the processing of music. NeuroImage. 2002;17(2):956–66.
93. Maess B, Koelsch S, Gunter TC, Friederici AD. Musical syntax is processed in Broca's area: an MEG study. Nat Neurosci. 2001;4(5):540–5. https://doi.org/10.1038/87502.
94. Patel AD. Language, music, syntax and the brain. Nat Neurosci. 2003;6(7):674–81. https://doi.org/10.1038/nn1082.
95. Norman-Haignere S, Kanwisher NG, McDermott JH. Distinct cortical pathways for music and speech revealed by hypothesis-free voxel decomposition. Neuron. 2015;88(6):1281–96. https://doi.org/10.1016/j.neuron.2015.11.035.
96. Chvojka J, Kudlacek J, Chang W-C, Novak O, Tomaska F, Otahal J, Jefferys JGR, Jiruska P. The role of interictal discharges in ictogenesis—a dynamical perspective. Epilepsy Behav. 2019;106:591. https://doi.org/10.1016/j.yebeh.2019.106591.
97. de Curtis M, Avanzini G. Interictal spikes in focal epileptogenesis. Prog Neurobiol. 2001;63(5):541–67. https://doi.org/10.1016/S0301-0082(00)00026-5.
98. Dichter M, Ayala G. Cellular mechanisms of epilepsy: a status report. Science. 1987;237(4811):157–64. https://doi.org/10.1126/science.3037700.
99. Jensen MS, Yaari Y. The relationship between interictal and ictal paroxysms in an in vitro model of focal hippocampal epilepsy. Ann Neurol. 1988;24(5):591–8. https://doi.org/10.1002/ana.410240502.
100. Wada JA, Sato M, Corcoran ME. Persistent seizure susceptibility and recurrent spontaneous seizures in kindled cats. Epilepsia. 1974;15(4):465–78. https://doi.org/10.1111/j.1528-1157.1974.tb04022.x.
101. Goncharova II, Alkawadri R, Gaspard N, Duckrow RB, Spencer DD, Hirsch LJ, Spencer SS, Zaveri HP. The relationship between seizures, interictal spikes and antiepileptic drugs. Clin Neurophysiol. 2016;127(9):3180–6. https://doi.org/10.1016/j.clinph.2016.05.014.
102. Karoly PJ, Freestone DR, Boston R, Grayden DB, Himes D, Leyde K, Seneviratne U, Berkovic S, O'Brien T, Cook MJ. Interictal spikes and epileptic seizures: their relationship and underlying rhythmicity. Brain. 2016;139(4):1066–78. https://doi.org/10.1093/brain/aww019.
103. Barbarosie M, Avoli M. CA3-driven hippocampal-entorhinal loop controls rather than sustains in vitro limbic seizures. J Neurosci. 1997;17(23):9308–14. https://doi.org/10.1523/JNEUROSCI.17-23-09308.1997.
104. Motalli R, Louvel J, Tancredi V, Kurcewicz I, Wan-Chow-Wah D, Pumain R, Avoli M. GABA$_B$ receptor activation promotes seizure activity in the juvenile rat hippocampus. J Neurophysiol. 1999;82(2):638–47. https://doi.org/10.1152/jn.1999.82.2.638.
105. Rauschecker JP, Scott SK. Maps and streams in the auditory cortex: nonhuman primates illuminate human speech processing. Nat Neurosci. 2009;12(6):718–24. https://doi.org/10.1038/nn.2331.
106. Sachs ME, Ellis RJ, Schlaug G, Loui P. Brain connectivity reflects human aesthetic responses to music. Soc Cogn Affect Neurosci. 2016;11(6):884–91. https://doi.org/10.1093/scan/nsw009.

Signs Made Flesh: Body, Improvisation, and Cognition Through Semiotics

9

Gabriele Marino and Vincenzo Santarcangelo

9.1   Introduction

Semiotics, the "the science of signs" according to etymology (from the ancient Greek σημεῖον, sēmeîon, "sign") [1, 2] has been struggling with music for decades, trying to apply its own conceptual grids to this subject matter [3, 4]. Music semiotics (musical semiotics or semiotics of music) has inherited the famed "textualism" affecting both its disciplinary sources: general semiotics (with its ideological focus on natural verbal language, at least in the tradition developed after Saussure [5]) and musicology (with its ideological focus on symbolic notation, to which music has been traditionally reduced). The discipline has always had issues in dealing with the proverbial ineffability of the musical *datum* (can we translate music into words?) and has indulged foundational questions that prevented, indeed, the development of a specific coherent branch: is music a language? Is it conceivable as a sign? Is musical meaning abstract or referential? Can we identify a minimum unit of musical meaning? etc.

Just like the other products of culture, music is not "a thing," a portion of the ontological world ready-made to be identified and analyzed: rather, music is what human beings, organized in historically situated communities, define as such. Hence, the conception of music changes diachronically (across time) and diatopically (across space). With the invention of sound recording, the rise of Modernist phonographic aesthetics (the electronic, concrete, and electroacoustic), and the increasing stratification of musical communication (mediated by audiovisual

G. Marino (✉)
Università di Torino, Torino, Italy
e-mail: gabriele.marino@unito.it

V. Santarcangelo
Politecnico di Torino, Torino, Italy
e-mail: vincenzo.santarcangelo@polito.it

B. Colombo (ed.), *The Musical Neurons*, Neurocultural Health and Wellbeing,
https://doi.org/10.1007/978-3-031-08132-3_9

technologies), it became clear that music was no longer an abstract matter idealistically translatable into signs on the score, but a multimodal matter rooted in sound [6].

In the following paragraphs, we will use semiotics as a "meta-" disciplinary device, with the aim to reconcile it with psychology and enable a mutual translation: we will employ the post-cognitivist paradigm of enactivism to give "conducted improvisation" (a type of structured, collective, musical improvisation) a theoretical framework (in other words, we will employ conducted improvisation as an exemplification of enactivism), and, conversely, we will employ conducted improvisation as a metaphor of enactivism, as something capable to make this paradigm better understandable.

9.2   Semiotics and the Body

The human body is the paradoxical, neglected subject matter par excellence. Its presence is apparently so obvious to us that its meaning would truly unfold only when it is obstructed (there is an "obtuse" meaning opposed to the "obvious" one [7]); namely, when meaning is being manipulated or put into question, so as to reveal the constructed dimension of what seems "natural." Barthes, among the leading figures of literary structuralism (developed in the wave of Saussure), was the first semiotic scholar to criticize a music scholarship aimed at deleting the bodily and the sensible from its metalanguage and, in the first place, epistemology; his essays on music were unsystematic but pioneering, ahead of their times. With the famous "grain of the voice" [8] and the distinction, inspired by Kristeva, between the "phenosong" (singing carrying linguistic meanings) and the "genosong" (singing meant as the vocalization of the corporeal datum), he opened the path to sound studies (timbre being another neglected subject matter in music studies). With the "somathemes" [9], minimum units of bodily meaning inscribed as implied gestures in Schumann's piano fantasies, he opened the path to musical embodiment.

Barthes' intuitions stood as hapaxes until semioticians gradually started to consider the body, in the wake of Merleau-Ponty's phenomenology of perception [10] as the pivot of experience [11–13] and, therefore, enactivism as a suitable model of cognition overcoming static categories and old dichotomies (body vs. mind, perception vs. interpretation, nature vs. culture, individual vs. environment) [14].

9.3   Enactivism

The origins of the enactive paradigm lie in Bruner's [15] proposal of a threefold mode of representation: the symbolic (based on language), the iconic (based on visual perception), and the enactive one (based on action; it is the kind of knowledge coming through movements, so that "the body shapes the mind" [16]). Enactivism represents a paradigm shift in the history of cognitive sciences. This approach, an alternative to the naturalistic one held by materialists and functionalists, is a multidisciplinary set of studies alternatively gathered under the name of "embodied cognition," developed around the anti-dualistic hypothesis that the mind is not an isolated system coinciding with the brain, or anyway implemented by it, but rather a complex object that must be investigated in its constitutive relations with the body and the – biological, social, and cultural – environment which the organism is situated in. In other words, in this perspective, cognition is regarded as a dynamic activity, rather than a faculty of the individual [17–21]. Unlike computational cognitive science, which is based on shared implicit and explicit premises, embodied cognition is better referred to as a research program with no clear key characteristics other than the tenet that computational cognitive science has failed to appreciate the body's relevance in cognitive processing, and that doing so necessitates a dramatic re-conceptualization of the nature of cognition and how it must be investigated.

Enactivism focuses on the contribution of bodily sensory-motor processes and environmental factors to the definition of cognition: namely, on the relations established by the agent with the surrounding space. Noë's work [22–25] aims at investigating notions such as "consciousness" and "perception" on the basis of a dynamic model of interaction involving not only the brain but also the body and its surroundings. According to this approach, "perception" is not an internal process based on the computational elaboration of information-stimuli deriving, in a static way, from the external environment, but is intrinsically connected to the explorative activities exercised by the body in motion. In other words, "cognition is not the representation of a pregiven world by a pregiven mind but is rather the enactment of a world and a mind on the basis of a history of the variety of actions that a being in the world performs" [19: 9]. This "memory" or this "history of past actions" is what has been called "body schema" [26], a notion employed in psychology to refer to the implicit and practical "body map" that makes it possible to use efficiently our body in action.

Noë and Gallagher's recovery of the notion of "body schema" [16, 22, 27] does nothing but confirm the obvious; it is not necessary to pay attention to one's body parts in order to use them efficiently. In the same way, a performance would be negatively affected if an expert performing a practical activity focused his attention on the bodily mechanic of the task, instead of participating in the activity as a whole. As an example, one might refer to the very different actions simultaneously implemented by a drummer in a very single measure – e.g., to kick the bass drum, to keep the beat on the hi-hat or on a cymbal, to hit the snare with the stick – and to the implied notion of "drum independence." A leader conducting a certain number of performers is a typical example of an expert engaging in a practical and embodied activity, whereas gestures are a typical example of embodied cognition [28, 29].

9.4   Enactive Activities

In Noë's most recent work [25], the biological activity of breastfeeding is said to be a paradigmatic example of "organized activity," i.e., an activity, primitive and natural, which is extended in time, becoming the arena for the exercise of attention, looking, listening, doing, and undergoing. Organized activities "emerge out" of the single activities and are not governed by the control of any individual. Finally, they

have a "function," social, biological, or personal, and are (at least potentially) pleasurable. To be organized, an activity must be marked, according to Noë, by six
features:

1. It must be primitive, basic, or biological. Breastfeeding is not the achievement of
   high culture but is rooted in our mammalian origins.
2. Despite being basic, breastfeeding requires exercise and the recourse to evolving
   and highly sophisticated cognitive skills on the part of both mothers and infants.
3. Since it is organized in time – since it begins, develops, and comes to an end – it
   has to be "structure." Notice that Noë significantly compares the structure of
   breastfeeding to that of turn-taking in conversation. Breastfeeding is really structured as a kind of "primitive conversation."
4. In breastfeeding, none of the subjects involved (neither mother nor infant)
   orchestrate or direct the activity they are involved in: "with its delicate interplay
   of listening and acting, doing and feeling, and with its distinctive turn-taking
   temporal dynamics, just sort of happens" [25: 4];
5. Breastfeeding has a "function": some of its aspects remain somewhat ambiguous
   and still unknown, but for sure it "must have something to do with feeding and
   with creating a relationship of attachment between mother and child" [25: 4].
   This kind of activity, thus, seems to be a peculiar kind of exercise, i.e., a
   relationship-building exercise.
6. Finally, albeit being an almost worklike source of conflict, a negotiation, breastfeeding is also a source of pleasure for both the subjects involved.

Let us consider another example proposed by Noë [25: 12–14]. We invite the
reader to compare it with the activity of musical improvisation, in order to discover
the grossly noticeable similarities and the potential differences. Dancing, according
to Noë, is an organized activity, if anything is.

1. Dancing is a spontaneous physical response to rhythm, to music, and to movement. Krueger [30] describes dancing as an embodied response to musical
   events, in which the temporal regularities of melodic and rhythmic musical patterns are embodied in a vast array of different bodily movements.
2. For these reasons, dancing is also an impressive exercise of powers of attention
   and perceptual discrimination.
3. Thirdly, dancing is clearly structured, since it is an activity organized in space
   and time, just like breastfeeding and conversation.
4. Albeit we dance on purpose, we enact a series of "expressive movements" (as
   opposed to "goal-directed movements"), which do not aim at realizing practical
   goals. In this kind of movement, we do not move "through" space, but "in"
   space. When we dance, we do not move just in order to reach a different point in
   the surrounding space: directions and distances lose the central role they have in
   goal-directed movements. The role of our movements is completely redefined by
   and subordinated to the expressive features that movements are planned to con-
   vey – by the "dance itself." In dancing, we are not interested in measuring the

space we move in, or – to say it with different words – our movement is not limited by points of departure and arrival: movement is not characterized by a pointed orientation in space. This is shown by the fact that in most cases dance-floors do not need to have specific shapes. Dancing is not influenced by the orientation or shape of the space in which it takes place because it is not a goal-directed kind of movement, with a place to leave and one to reach: consequently, dancers do not decide how to dance, at least not at the level of the way their movements are swept up into and organized by the dancing. The dancing just occurs; even if one dancer may "lead," this is just a special way of letting oneself be caught up in the dance: "a good dancer is in the flow." [25: 12].

5. Even lacking the practical background of approaching and moving away – that is, the system of goals and directions, points of departure and points of arrival, distances and orientation – dancing is organized on the basis of a precise system of meanings, that is, the one of "expressivity." We might dance to express our feelings, or to establish ourselves as having an identity. Movements in dancing are not oriented at achieving a practical goal, but at expressing affective valences. More generally, "dancing has a point. Some people dance to meet girls or boys. Sometimes we dance […] because this is demanded of the situation" [25: 12].

6. And finally, being at once basic and spontaneous while also cognitively sophisticated, dancing is, or at least can be, a pleasurable activity.

9.5   Conducted Improvisation

"Conducted improvisation" (a calque from 31; in literature one can also find "controlled," "structured," or "composed" improvisation) is the overextended category we propose to designate a form of organized, collective, musical improvisation wherein the figure of a "conductor," who delivers instructions to the performers, mainly using gestures and graphic scores, is established.

The main difference between simple "collective improvisation" (e.g., Coleman's *Free jazz*, 1961) and conducted improvisation (which may be considered as a particular type of the first category) lies in the systematic nature of the latter. Conducted improvisation enduringly employs a specific and shared lexicon, through which codified ways of interactions between the involved subjects (i.e., the conductor and the ensemble, the conductor and one musician, the musicians themselves both as singles and as part of sub-groups in the ensemble) are established. Feedbacks (i.e., the performer's acceptance or refusal of the instruction delivered by the conductor) play a key role in the construction of the performance.

A provisional outline of conducted improvisation throughout history might include: Russolo's noise intoners orchestra, Stockhausen's Intuitive Musik, Cage's event music, Brown's open form, Wolff's cues and game pieces, Xenakis' stratégie musicale, Sun Ra's Arkestra performances, Davis' silent way, Zappa's Mothers of Invention musical theatre, Eno's oblique strategies, Thompson's Soundpainting, Morris' Conduction®, and Zorn's file card and game pieces. The two latter cases represent the most systematic and documented examples of conducted improvisation.

9.5.1   Morris' Conduction

Drawing inspiration from musicians who had enduringly worked with ensembles in a workshop-like fashion [32: 2], Butch Morris (1947–2013) started to develop a method for composing improvisations live, online, in the 1970s, but the first public performance of what he had called "Conduction" (a portmanteau word – explicitly modeled upon the homograph from physics – pinched with a deconstructionist flavor, made up with "conducting" and "improvisation") would have taken place in 1985 (published in 1986 with the title *Current trends in racism in modern America*).

Morris, who started his career as a jazz cornetist with bandleader David Murray, devoted most of his life to the worldwide diffusion – through 199 accounted workshops/performances, involving musicians he had never met before – of his method. The gestures he employed, an expansion of traditional conducting, constituted a codified and coherent lexicon, by means of which he intended to join the traditions of European classical music and Afro-American jazz together [33]. It is worth reading Morris' official definition of Conduction:

> Conduction (conducted Improvisation) is a means by which a conductor may compose, (re) orchestrate, (re)arrange and sculpt with notated and nonnotated music. Using a vocabulary of signs and gestures, many within the general glossary of traditional conducting, the conductor may alter or initiate rhythm, melody, harmony, not to exclude the development of form/structure, both extended and common, and the instantaneous change in articulation, phrasing, and meter. Indefinite repeats of a phrase or measures may now be at the discretion of the new Composer on the Podium. Signs such as memory may be utilized to recall a particular moment and Literal Movement is a gesture used as a real-time graphic notation. Conducting is no longer a mere method for an interpretation but a viable connection to the process of composition, and the process itself. The act of Conduction is a vocabulary for the improvising ensemble [32: 5].

9.5.2   Zorn's *Cobra*

John Zorn (b. 1953), who participated as a saxophonist in Morris' first issued Conduction, describes himself as a contemporary composer struggling with the paradox of having to write music for improvisers, since that of the radical New York improvisers (at the intersection between free jazz and free improvisation, or non-idiomatic improvisation) was the natural context of his musical *Bildung* as well as his ordinary working environment. Like Morris, Zorn explicitly lists his influences [34] and focuses on the interpenetration between composition and improvisation; what he calls "game pieces" are nothing but the programmatic exploration of this way of music-making: systems of rules, gestures (including usage of paraphernalia), graphic indications, and roles are designed so as to structure the performances of the improvisers through the figure of the "prompter."

Zorn always tries to insert an improvisational moment in compositions that, otherwise, would be entirely notated and, conversely, always finds ways to regiment improvisation; he claims that every note of his music, even the improvised one, must be reasoned, significant, and important, hence the mixture of the two modalities. *Cobra*, named after a simulation game set during the Second World War, developed in 1984 and released for the first time in 1987 (with recordings made between 1985 and 1986), represents the sum of the work on the game piece format experimented by Zorn in the 1970s.

9.6   Comprovisation

The semiotic square is a classification device elaborated by Greimas (the leader of the structural-generative approach to semiotics) and perfected and popularized by his pupil Floch that derives from classic logic (Aristotle), providing the visualization of a given semantic category [5: 308–311]. The semantic category designed through a semiotic square is identified not only the contradictory terms (A vs. non-A), but also the contrary ones (A vs. B). By building a semiotic square we may map the axiologies (valorizations) at stake in a given field of human experience.

Being the composition of an improvisation (Morris also employed the term "comprovisation"), conducted improvisation stands as the complex term of the opposition "composition vs. improvisation" (the contrary terms at the basis of the consequent semiotic square), deconstructing both habitual contexts of music playing, their organizational models, and underlying values. Conducted improvisation builds up a type of performance and a type of environment that is challenging for all the subjects involved in the process: the performers have to learn entire sets of body schemas, which are completely new to them, in a short term (during the workshops preceding the on-stage performance); the conductor has to consider the feedbacks coming from the performers, in order to deliver the subsequent instruction. In this perspective, a circular feedback circuit is established; the environment affects the direction and the direction manipulates the environment.

A typical Gibsonian concept [18, 21], that of "environment," is employed by Morris to describe his musical practice as the organization of the surrounding things, conditions, and influences. Morris meant Conduction as the "art of environing," an act of intersemiotic translation capable of turning the "character of the environment" [32: 4] into sound: he wanted to turn the actual place where he was working and the actual musicians with whom he was working into music.

In Morris' Conduction, the musicians follow the movements of the baton, of the hands, and, in general, of the body of the conductor; their way of receiving and interpreting the instructions inscribed in these movements influences, in turn, the subsequent ones of the conductor, who can confirm or contradict those interpretations, developing the paths suggested by the musicians or, on the contrary, suggesting different ones. In Zorn's *Cobra*, the "conversational turns" are regulated by the prompter through three types of signals: hand movements (hand signals and deictic gestures, such as indicating parts of the body or the musicians); colored signs with letters and symbols; using a hat (the prompter may wear it or not, and wave it). The one musician convoked by the prompter may or may not accept the task assigned and, therefore, may force the prompter to delegate it to others.

In both Morris and Zorn's conducted improvisations, the conductor/prompter does not take part in the musical performance by playing an instrument, but still is present as a performer on-stage, generally at the center, in a "teacher in the classroom"-like fashion.

9.7 Enactive Improvisation

It is possible to set conducted improvisation within the enactive paradigm, in two ways: by labeling this form of musical performance as an enactment-driven practice; by defining it as a metaphor (properly, a prosopopoeia; personification, in rhetoric) of the enactive processes themselves.

The "lexicon" of Conduction (formerly, "vocabulary"; an abstract of which is available in 32: 6–7) is being systematically studied by Veronesi [35–37], a linguist who had also collaborated with Morris as an interpreter during his Italian residencies. Veronesi backs a pragmatic perspective, with the aim to enlighten the multimodal features of this practice.

Conduction, indeed, is a musical and performative practice wherein various semiotic resources (talk, gestural imitation of instrumentalists' actions, vocal exemplifications, verbal and bodily enactments of directive sequences) are "laminated" (or "simultaneously layered" [38, 39]) and elaborate each other [37]. Therefore, Conduction employs a set of "gestural metaphors and metonymies" [35], which may be understood as "metaforms" (any form that connects two different domains, like an abstract notion to a concrete source, as in the case of metaphors [40]), or "plastic formants" (basic unit of non-figurative visual semiotics [41]; each single, recognizable, meaningful gesture may be understood as a gestural plastic formant, reminiscent of Barthes' somatheme).

It is worth quoting the complete description of a typical Conduction instruction and visualizing it (Fig. 9.1):

Fig. 9.1 Butch Morris performing the instruction "Expand" (or "Develop") from the Conduction lexicon. Graphic reworking of a photograph taken from [39: 98]

DEVELOP. Description of Gesture: Hands palm to palm facing left and right, chest level, separating left and right and returning. Meaning: Is used to variate, elaborate, embellish, transform, adorn, manipulate, augment, diminish, fragment, deconstruct or reconstruct a specific "point of information." Explanation: When the palms are together, this is the position of the specific information (idea or point of information) to be developed. As the hands separate the development of information takes place, as the hands return to the together position a reconstruction of the idea takes place, when the hands reach the together position this is the downbeat for the return to the initial information. The degree of development is determined in stages by the space between the hands. [42: 178].

By "actant" semiotics defines any syntactic position occupied by a given agent, human or non-human, within a given story and, therefore, text (any possible object of analysis for semiotics); e.g., the main Subject, meant as the Hero, of the story. Actants are roles or functions and must be distinguished from "actors" (characters of the story, to put it simple; e.g., the main Subject or Hero, in this particular story, is the Prince Whatchamacallit). The very same actantial role may be portrayed by different "actors" (many characters may serve as Helpers of the Hero) and, vice versa, the very same actor may carry different actantial roles (a character may serve as Helper in the beginning and turn out to be an Opponent afterward) [5: 5–9].

As a matter of fact, Conduction provides the actantial positions implied – and, normally, un-staged – in musical improvisation (and in musical performance in general) with physical actors; here lies its metaphorical value for enactive cognition. In other words, the conductor, delivering the instructions to the performers, does embody and makes the constraints that are working underneath the musical practice (e.g., architextual, stylistic, and conversational norms) visible.

9.8 Conclusion

By explicitly showing the existence of rules, the asymmetry and fragility of relationships, these practices stage the "behind the scenes" of musical improvisation – we can think of them as a form of Ur-Improvisation – and of musical performance in general, stressing the intersubjective and contractual character of cognition and signification (meaning-making). Morris and Zorn show on-stage the music-making, even though they do not necessarily grant the audience full access to it: they do show the elements of the code (the signals), not the key to it (the rules, the meta-signs, the metalanguage). They are interested not in showing a static result, but rather a dynamic process, not in producing enunciates, but rather enacting the enunciation, not making the listener hear "played music," but rather music-in-the-making; just like Cézanne, with his obsessive visual research on Mont Sainte-Victoire, "wanted to depict matter as it takes on form" [43: 13], according to Merleau-Ponty. Conducted improvisation is the staging, the enactment of enaction itself (of the embodiment of musical knowledge).

The enactive paradigm is being increasingly employed as a theoretical framework for dealing with aesthetical objects, including music [24, 44–47], also in an ecological semiotic perspective [48–51]. Due to its circular autopoietic nature [52] and its cooperative and didactical component [36], as it shifts the focus of music-making from the organization of sound to the organization of musicians, conducted improvisation may find a promising field of application in educational, re-educational, rehabilitational, and music therapical contexts. It is no coincidence that Zorn himself has defined his practice as a "psychodrama," thus reconnecting it to Globokar's "catalog of reactions that we can prescribe to an interpreter."

Acknowledgments The research presented herein is the result of close cooperation between the authors. However, for formal attribution, please consider par. 9.1, 9.2, 9.5 and 9.7 as authored by Marino, and par. 9.3, 9.4, 9.6 and 9.8 by Santarcangelo. The authors wish to thank Dr. Antonino Giordano and Dr. Prof. Bruno Colombo.

References

1. Nöth W. Handbook of semiotics. 2nd ed. Bloomington IN: Indiana University Press; 1995.
2. Chandler D. Semiotics: the basics. 4th ed. London-New York: Routledge; 2022.
3. Monelle R. Linguistics and semiotics in music. Newark NJ: Harwoord Academic Publishers; 1992.
4. Tarasti E. Signs of music: a guide to musical semiotics. Berlin-London: de Gruyter; 2002.
5. Greimas AJ, Courtés J, editors. Semiotics and language: an analytical dictionary. Bloomington IN: Indiana University Press; 1982. [or. ed. 1979].
6. Kania A. The philosophy of music. In: Zalta EN, editor. The Stanford encyclopedia of philosophy (Fall 2017 Edition). https://plato.stanford.edu/archives/fall2017/entries/music/.
7. Barthes R. Le troisième sens: notes de recherches sur quelques photogrammes de S. M Eisenstein. Cahiers du cinema. 1970;222:12–9.
8. Barthes R. Le grain de la voix. Musique en jeu. 1972;9:57–63.
9. Barthes R. Rasch. In: Kristeva J, Miller JA, Ruwet N, editors. Langue, discours, société. Pour Émile Benveniste. Paris: Seuil; 1975. p. 218–28.
10. Merleau-Ponty M. Phenomenology of perception. New York: Routledge and Kegan Paul; 1962. [original French edition 1945].
11. Fontanille J. Soma et séma: Figures du corps. Paris: Maisonneuve et Larose; 2004.
12. Marrone G. Introduction to the semiotics of the text. Berlin-Boston: de Gruyter; 2022.
13. Finol JE. On the corposphere. Anthroposemiotics of the body. Berlin-Boston: de Gruyter; 2021.
14. Paolucci C. Cognitive semiotics. Integrating signs, minds, meaning and cognition. Cham, Switzerland: Springer; 2020.
15. Bruner J. Toward a theory of instruction. Cambridge MA: Belkapp Press; 1966.
16. Gallagher S. How the body shapes the mind. Oxford: Oxford University Press; 2005.
17. Bateson G. Steps to an ecology of mind. San Francisco: Chandler; 1972.
18. Gibson JJ. The ecological approach to visual perception. Boston: Houghton Mifflin; 1979.
19. Varela FJ, Thompson E, Rosch E. The embodied mind: cognitive science and human experience. Cambridge MA: MIT Press; 1991.
20. Clark A, Chalmers D. The extended mind. Analysis. 1998;58(1):10–23.
21. Santarcangelo V. Introduzione. In: Gibson JJ, editor. L'approccio ecologico alla percezione visiva. Mimesis: Milano-Udine; 2016. p. IX–XXI.
22. Noë A. Action in perception. Cambridge MA: MIT Press; 2004.
23. Noë A. Out of our heads: why you are not your brain, and other lessons from the biology of consciousness. New York: Hill & Wang; 2009.
24. Noë A. Varieties of presence. Cambridge MA: Harvard University Press; 2012.
25. Noë A. Strange tools. Art and human nature. New York City: Hill and Wang; 2015.
26. Head H, Holmes G. Sensory disturbances from cerebral lesions. Brain. 1911;34(2–3):102–254.
27. Gallagher S, Cole J. Body schema and body image in a deafferented subject. J Mind Behav. 1995;16(4):369–90.

28. Kendon A. A description of a deaf-mute sign language from the Enga Province of Papua New Guinea with some comparative discussion. Semiotica. 1980;31(1–2):1–34.
29. Streeck J. Gesturecraft: The manu-facture of meaning. Amsterdam: John Benjamins; 2009.
30. Krueger J. Enacting musical experience. J Conscious Stud. 2009;16(2–3):98–123.
31. Salvatore G. Work in progress: il teatro musicale dei Mothers of Invention (1967–69). In: Id, editor. Frank Zappa domani: Sussidiario per le scuole (meno) elementari. Roma: Castelvecchi; 2000. p. 47–61.
32. Graubard A, Morris B. Liner notes of Butch Morris, testament: a conduction collection, New World, CAT 80478–2; 1995. https://bit.ly/3pYckUR (via InternetArchive).
33. Veronesi D, editor. Morris L.D.B. The art of conduction: a conduction workbook. New York: Karma; 2017.
34. Brackett J. Some notes on John Zorn's cobra. Am Music. 2010;28(1):44–75.
35. Veronesi D. The guy directing traffic: gestures in conducted improvised music between metaphor and metonymy. Paper presented at the RaAM Workshop Metaphor, Metonymy & Multimodality, Amsterdam, June 4–5; 2009.
36. Veronesi D. La traduzione non professionale come co-costruzione: osservazioni sull'interazione in contesti musicali didattici all'intersezione tra codici semiotici. In: Massariello MG, Dal MS, editors. I luoghi della traduzione: le interfacce. Proceedings of the XLIII international congress of SLI-Italian linguistic society, vol. I. Roma: Bulzoni; 2011. p. 89–102.
37. Veronesi D. Formulating music, laminating action: instruction and correction in ensemble music workshops. Paper presented at the Linguistic Anthropology Laboratory, University of California, San Diego, May 21; 2012.
38. Goffman E. Forms of talk. Philadelphia: University of Pennsylvania Press; 1981.
39. Goodwin C, Duranti A. Rethinking context: an introduction. In: Id, editor. Rethinking context: language as an interactive phenomenon. Cambridge MA: Cambridge University Press; 1992. p. 1–42.
40. Danesi M, Sebeok TA. The forms of meaning: modeling systems theory and semiotic analysis. Berlin-London: de Gruyter; 2000.
41. Greimas AJ. Sémiotique figurative e sémiotique plastique. Actes Sémiotiques. 1984;VI(60):5–24.
42. Stanley TT. Butch Morris and the art of Conduction®. PhD thesis, University of Maryland, US; 2009. https://drum.lib.umd.edu/handle/1903/9935.
43. Merleau-Ponty M. Sense and non-sense. Evanston IL: Northwestern University Press; 1964.
44. Luciani A. Enaction and music: anticipating a new alliance between dynamic instrumental arts. J New Music Res. 2009;38(3):211–4.
45. Peters D. Enactment in listening: intermedial dance in EGM sonic scenarios and the bodily grounding of the listening experience. Perform Res. 2010;15(3):81–7.
46. Matyja JK. Review of Clarke D., Clarke, E. (eds.). Music and consciousness: philosophical, psychological, and cultural perspectives. New York: Oxford University Press; 2011. Constr Found. 2012;8(1):129–31.
47. Lopez-Cano R. Music, cognitive enaction and social Metabody. Paper presented at Multiplicities in motion: Affects, embodiment and the reversal of cybernetics, Medialab Prado, Madrid, July 27; 2013.
48. Reybrouck M. Biological roots of musical epistemology: functional cycles, umwelt, and enactive listening. Semiotica. 2001;134(1–4):599–633.
49. Reybrouck M. Music knowledge construction: enactive, ecological, and biosemiotic claims. In: Lesaffre M, Maes PJ, Leman M, editors. The Routledge companion to embodied music interaction. London: Routlegde; 2017. p. 58–66.
50. Reybrouck M. Music as experience: musical sense-making between step-by-step processing and synoptic overview. In: Sheinberg E, Dougherty WP, editors. The Routledge handbook of music signification. London: Routledge; 2020. p. 277–84.
51. Reybrouck M. Musical sense-making. Enaction, experience, and computation. Abingdon-New York: Routledge; 2021.
52. Maturana HR, Varela FJ. Autopoiesis and cognition: The realization of the living. Dordecht: D. Reidel Publishing Co.; 1980.

Flute and Mind

10

Lorenzo Lorusso, Francesco Brigo,
Antonia Framcesca Franchini, and Alessandro Porro

10.1   Introduction

Many scholars, including some philosophers, have been fascinated by the relationship between music and the brain/mind [1, 2]. Nowadays, thanks to the progress achieved in the field of neuroscience, it's possible to investigate the neural correlates of music.

Among musical instruments, the flute is one of the most ancient [3] and has been widely used in different musical cultures around the world. Furthermore, its extreme versatility has contributed to its popularity in different musical genres, from classic music to rock music, and from military marches to bossa nova [4].

In this chapter, we provide an overview of the historical evolution of this instrument in western culture. Furthermore, we discuss how in its century-long history, the flute has been used to express the intimate relationship with the mind and its derangement.

10.2   Historical Evolution of the Flute

The flute is a family musical instrument in the woodwind group, which dates back to the paleolithic period with important transformation in the following centuries in shape (transverse flute and vertical/end-blown (Figs. 10.1 and 10.2), in playing style

L. Lorusso (✉)
UOC Neurologia & Stroke Unit, A.S.S.T.-Lecco, Dipartimento di Neuroscienze,
P.O. Merate (LC), Italy

F. Brigo
Department of Neurology, Hospital of Merano-Meran (SABES-ASDAA), Merano, Italy

A. F. Franchini · A. Porro
Dipartimento di Scienze Cliniche e di Comunità e CRC Centro di Salute Ambientale,
Università degli Studi di Milano, Milan, Italy

© The Author(s), under exclusive license to Springer Nature
Switzerland AG 2022
B. Colombo (ed.), *The Musical Neurons*, Neurocultural Health and Wellbeing,
https://doi.org/10.1007/978-3-031-08132-3_10

Fig. 10.1 Allegory of Music (ca 1600), Van Ravesteyn Dirck de Quade (1576–1612). Kunsthistorisches Museum, Vienna. A noblewoman transverse flute player (public domain)

Fig. 10.2 Jacob Jordaens (1593–1678). Three strolling musicians (1645–1650). Museo del Prado, Madrid. The first plays a vertical or end-blown flute while the others two sing from sheet music held by one of them (public domain)

and repertoire, in the lives of flute players and makers, and in the uses of the instrument to play military, religious, consort, solo, chamber, opera, symphony, jazz, popular, and flute band music [5].

10.2.1 The Flute in the Paleolithic Period

The musical language was possibly born before the verbal language. As already suggested by Charles Darwin (1809–1882), language could have emerged from call vocalizations as first attempts of communication between human beings, similarly to what occurs among other animals [6]. According to the neuroscientist and neuropsychologist Jaak Pankseep (1943–2017), music derives from the cries issued from the first hominids when some of them walked away from the group [7]. Among animals, cries can preserve the contact between mothers and puppies; when the puppy hears the voice of the mother, its fur rises and warms the animal up. Music is therefore associated with changes in the vegetative system, supporting its biological significance. No other means of communication can evoke equally strong emotional reactions. It is, therefore, reasonable to hypothesize that melodic instruments have appeared before percussion instruments, although the latter might appear easier to construct and hence more primitive. The oldest musical instrument identified as such is the flute of Divje Babe (named after the site of its discovery), obtained from the femur of a cub. It dates to the Paleolithic period and is probably 43,100 years old. It is possible, however, that the first instruments date back to even 50,000–60,000 years ago [8].

10.2.2 The Flute in Ancient Greece and Rome

Historical evidence emphasizes the importance of the flute in ancient Greece and Rome. It was used in tragedies and in the elegy, a literary genre that was named after the Greek term indicating the flute (*élegos*), the instrument that accompanied the declamation of poems. Furthermore, the flute and the drum were used to mark time during military marches, a tradition that has lasted over the centuries. In ancient Greece and Rome, there were two types of flute [9]. The first was the Greek wind instrument called *aulòs* (in Latin *tibia*, named after the bone with which it was built, although there were instruments made of wood, sugarcane, or ivory). It consisted of a tube with a bulb mouthpiece and a reed. The version with two diverging tubes was called *dìaulos*. According to the myth [10], this instrument was invented by Athena. The goddess threw it away, after realizing that it caused her cheeks to puff off, ruining her beauty. The *aulòs* was then picked up by the satyr Marsia, who became so good at playing it that he challenged Apollo, the god of music, confident that he would beat him. The god accepted and asked the Muses to judge the competition. At first, the jury was very impressed by the melodies of Marsia's flute. Apollo – fearing for his defeat – started playing his lyre, singing at the same time, and challenged his rival to do the same. Marsia was unable to do so, and Apollo won the contest. As punishment for challenging a god, Apollo inflicted dreadful torture on Marsia: tied to a tree, he skinned him alive. The second type of instrument was the Pan flute (also called syringe or Pan-pipes), composed of multiple reeds of different lengths, usually joined together, that was played by blowing through the upper openings of the reeds.

10.2.3 The Flute from Medieval Times to Nowadays

As already mentioned, the flute was present in the Italian peninsula and the Mediterranean basin since antiquity, as demonstrated by numerous findings and iconographic sources. However, in the early Middle Ages, after the barbarian invasions, the flute almost disappeared. The advent of Christianity, which preferred vocal-choral music over instrumental music (associated with pagan culture), was also partly responsible. It has been hypothesized that the flute was reintroduced in Europe by populations involved in pastoralism, coming from the Byzantine Empire and the Balkan peninsula [11]. From the middle ages, the distinction between transverse flute and vertical (end-blown) flute became increasingly evident, with different historical evolution, musical repertoire, and playing techniques. Although both belong to the subclass of aerophones, it would be inappropriate to evaluate together their historical evolution. In this chapter, we have focused on the historical evolution from Medieval to present times of transverse flute alone. The first recognized evidence of the transverse flute in the Middle ages is an image from the *Hortus deliciarum* ("Garden of delights"), a manuscript found in the Alsatian monastery of Hohenbourg dating between 1159 and 1175 AD [12]. The image depicts an allegory of temptations, where the flute is associated with vainglory, singing with avarice, and harp with lust. In the description, the instrument is named not only with the Latin term *tibia*, but also with the German name *Swegel*; the vernacular term suggests that the instrument was present in the transalpine territory and, therefore, the manuscript was not a copy of prior Byzantine oeuvres. In this and other surviving iconographic representations, the medieval flute was played with both hands and had aligned holes, like the models used in the Renaissance. The shape of the Renaissance flute (from the early to the mid-sixteenth century, also covering the first Baroque period) reflected the mentality and taste of the time: generally sober and linear, it had no additional ornaments and moldings. The instrument could be heard on various occasions in the cities and at court, but also during the war: the Lansquenets troops marched at the rhythm of the *Spiel* (from the German *Spielen*, to play), the sound of flute and drum [12]. At the English court, the flute was widely used, and king Henrich VIII (1491–1547) played the end-blown flute and the virginal, singing, dancing, and composing polyphonic elaboration of arias and ballads [11, 13]. It was thanks to the increasing cultural exchanges among the European courts that the flute reappeared in Italy around 1520–1530.

In the eighteenth century, Johann Joachim Quantz (1697–1773) (Fig. 10.3) composed several flute sonatas and about 300 concerts, and his flute method was highly influential [11]. He also played for Frederick II of Prussia (Frederick the Great, 1712–1786), who invited him to become his court composer and personal teacher; the king himself became a remarkable flutist, who also wrote a method for flute [12].

Other important composers of the eighteenth century who created musical pieces for this instrument were Antonio Vivaldi (1678–1741), Alessandro Scarlatti (1660–1725), Georg Philipp Telemann (1681–1767), Carl Philipp Emanuel Bach (1714–1788), and Johann Christian Bach (1735–1782) [13, 14]. In the eighteenth and early-nineteenth centuries, the mechanics of the flute was changed and keys were added to make it sound more uniform and to simplify some awkward fingerings. These technical improvements contributed to a renewed interest of numerous composers toward this instrument, especially Ludwig van Beethoven (1770–1827) and Wolfgang Amadeus Mozart (1756–1791) (particularly remarkable is his opera,

Fig. 10.3 Johann Joachim Quantz (1697–1773), portrait by Johann Friedrich Gerhard, 1735. Bayreuth, Neues Schloss, Musikzimmer (public domain)

Die Zauberflöte, "the Magic Flute," 1791). The flute was one of the first musical instruments used in opera [15].

However, it was during Romanticism that the flute reached its greatest success, especially thanks to Theobald Böhm (1794–1881), a flutist virtuoso, German inventor, and composer, who further modified the flute. He made small adjustments in the shape of the instrument and developed a system of keys that led to a more accurate intonation of octaves, improved ergonomics, and optimal acoustics [11, 12, 16]. The result was an instrument that was much easier to play and with a wider range than prior traditional flutes. This modern flute caught the attention of composers like Claude Debussy (1862–1918; among his compositions for flute solo: *Syrinx*, 1913) and Maurice Ravel (1875–1937) (the flute has a remarkable role in the ballet Daphnis et Chloé, 1912).

In the twentieth century, the flute proved its great versatility and established itself as one of the protagonists of the modern orchestra. It has been increasingly used also in new musical genres, such as jazz, where it can produce more sweet and atmospheric sounds through conventional emission or more aggressive sounds obtained through the humming technique. The latter consists in singing the same note on the flute, or a different one from that played, instead of simply blowing inside the instrument. This technique was created in the 1950s by the American flutist Sam Most (1930–2013) in a rather curious way: since he lived in an apartment building, most was allowed to play only for a few hours per day. Hence, to train without disturbing his neighbors, he used to play with a very low volume. In doing so, he realized that he was singing in the flute rather than blowing in it [17]. Since then, this technique has been used by many jazz flutists, like Ronald Kirk (1936–1953) and James Newton (1953–), and even in rock music, where it was made famous by Ian Anderson (1945–), frontman of the rock band Jethro Tull [12].

10.3 The Neural Correlates of Flute Playing

Learning how to play the flute requires the acquisition of numerous neuropsychological and neuromotor skills, which develop the function of various brain regions, including the auditory areas, involved in the recognition and processing of harmony and rhythm, and the somatosensory motor areas [18]. The latter play a role in maintaining a determined posture, associating each note to a specific fingering (i.e., a specific position of the fingers on the flute), modulating the intensity of blowing, and controlling the breathing. Playing the flute requires a complex interplay between the muscles of the hands, the eyes, the tongue, the larynx, and the respiratory system to ensure the effective coordination of these anatomical structures [19–21].

During a musical performance, the motor systems control the movements needed to produce the sound that, in turn, is processed by the auditory system, eventually regulating the motor output. Such feedback process, similar to that occurring while singing, ensures a refined control during the performance, enabling an almost instantaneous and automatic adjustment of the sound [22]. While playing, the flutist aims to communicate emotions to the public, not through bodily movements, such as a mime or a dancer would do, but diverting and translating affections into instrumental technical gestures.

The skills required to play the flute also lead to a particular approach to listening to music. Compared to untrained subjects, musicians not only enjoy the sounds but tend to analyze the structure of the musical piece in its various rhythmic, harmonic, and tonal aspects, transforming the auditive experience into a more complex and all-encompassing experience [23].

10.4 The Flute and the "Sound of the Mind"

Like speaking and singing, playing the flute requires breathing. Thus, the flutists are deeply involved in the generation of the sound, which emanates from within their bodies, contributing to its fascination and with listeners best able to discriminate flute tones from other musical instruments [24]. From a historical perspective, it appears evident that these characteristics have been recognized since ancient times.

In his opera *Die Zauberflöte* (The Magic Flute, 1791), Wolfgang Amadeus Mozart (1756–1791) employed the sound of the flute as an instrument allowing the self-fulfillment of the main characters, the noble Tamino and Pamina, and their lower counterparts Papageno and Papagena. In the final scene of the first act, thanks to the sound of his magic flute, the prince Tamino – like a fairytale Orpheus – enchants the animals that come to listen to the music (aria: *Wie stark ist nicht dein*

Zauberton, "How strong is thy magic tone"). Then, using Papageno's pipe sounding offstage as a sort of musical compass, and dialoguing with it using his own flute, Tamino is eventually able to find his servant and the princess Pamina. According to this opera imbued with masonic elements, the flute becomes for human beings like a light in the darkness, a guide in the labyrinth of unbridled passions.

In ancient Greek mythology, the flute was associated with Pan, the god of the wilderness, fertility, and sexuality. The opening flute solo in *Prélude à l'après-midi d'un faune* by Claude Debussy (1862–1918) was meant to represent the wakening of a young faun from his noontime nap. In this symphonic poem, which is based upon the poem *L'après-midi d'un faune* ("The Afternoon of a Faun," 1867) by the French author Stéphane Mallarmé (1842–1898), the music reflects the sensual experiences of the faun and his encounters with several nymphs in a dreamlike fashion. Of note, Faunus was the Pan's counterpart in ancient Roman mythology, and the term "panic" refers to the shout that the god Pan would have given when accidentally awakened from his noontime sleep, causing flocks to stampede. Accordingly, Mallarmé chose to set the awakening of the faun from his afternoon sleep. The awakening of the faun represents therefore the awakening of the primeval forces of sexuality, as further shown in its first choreography: Debussy's work was choreographed by Vaslav Nijinsky (1889–1950) for the Ballets Russes, with Nijinsky himself dancing the role of the faun; his highly eroticized choreography caused a huge scandal at the première (Paris, 1912).

The potential for virtuosity and the wide range of the flute was employed to musically represent the most passionate love, a feeling that – for its intensity – is what most resembles psychosis (albeit, admittedly, a rather benign one). One can find an example of this use in the aria *Ah! perché la conobbi?/Invan strappar dal core* sung by Lord Sidney in the opera *Il viaggio a Reims, ossia L'albergo del giglio d'oro* ("The Journey to Reims, or The Hotel of the Golden Fleur-de-lis," 1825) by Gioachino Rossini (1792–1868). In this musical piece and the following cabaletta, the traverse flute translates into sounds the love and fiery passion of the otherwise reserved English colonel for the beloved Corinna.

The flute was traditionally used to mark the time during military marches (a famous example is *The British Grenadiers*, played by a flute ensemble and a bass drum). Thus, from early on, it was recognized as an instrument that could create a sense of belonging and coordinated movements in a large group of people. This use of the flute might have served as a basis for the tale of the Pied Piper of Hamelin (German: *der Rattenfänger von Hameln*), which was included by Jacob (1785–1863) and Wilhelm (1786–1859) Grimm, known as the Brothers Grimm, in their collection *Deutsche Sagen* (first published in 1816). The tale tells of a piper who offers to free the city from the rats in exchange for a reward. Thanks to the sound of his flute, he manages to lure the animals into a river where they drown. The mayor refuses to pay him the reward, provoking the wrath of Piper who enchants with his instrument all the children of the town, imprisoning them in a cave. Beyond its fairytale atmosphere, this legend shows how, according to the popular imagination, the sound of the flute was able to influence the mind and the behavior of animals and humans alike, exerting a sort of "mind control."

During the romantic period, the flute was also employed to represent madness as the most extreme form of derangement of the mind. The most famous example of this use appears in the mad scene from the opera "Lucia di Lammermoor" (1835) by Gaetano Donizetti (1797–1868). Lucia – who on her wedding night has just stabbed her new husband in the bridal chamber – appears on stage, in a trance-like state, recalling her prior meetings with the beloved Edgardo and imagining herself married to him. In this scene, the vocal part is deeply intertwined and echoed by the sounds of the flute. However, originally Donizetti intended the voice of Lucia to be accompanied by the sound of the glass harmonica [25]. Interestingly, this instrument – invented by Benjamin Franklin (1706–1790) in 1761 – had been used by the German physician Franz Anton Mesmer (1734–1815) to induce a state of trance among patients suffering from various disorders [26]. Furthermore, glass harmonica was popularly associated with melancholia – a condition characterized by depressed mood, hallucinations, and delusions. This association probably arose because harmonica players suffered lead poisoning caused by the lead in rotating glass bowls [26, 27]. Despite being originally conceived for glass harmonica, Donizetti was forced to replace it with the flute, as the glass-harmonica player in Naples stepped out short before the première [28]. Since then and until very recently, the voice has been accompanied by the fascinating, although more conventional, timbre of the flute. The potential of the flute to express mind derangement and hallucinations has been explored also in contemporary music. An example of this use can be found in the solo flute piece *Cassandra's Dream Song* (1970), by the English composer Brian Ferneyhough (1943–). The outstanding complexity of this musical piece redefines and pushes flute playing to its extreme technical and expressive limits.

10.5 Conclusions

The intimate connection between the sound of the flute and the "sound of the mind" has a profound historical origin and popular foundation. Music can affect the functioning of our brain, leading to emotional and motor reactions. With its fascinating and long-lasting history, the flute – an instrument so akin to the human voice – continues to exert a deep influence on the human brain.

Acknowledgments This chapter is dedicated to Sebastiano Chia (18.08.1996–13.08.2015) who loved the flute and music. He presented a thesis on the relationship between flute and mind as a high school graduation thesis playing some musical pieces on the flute. His passion involved all those who knew him and set up this chapter that he had no way of being able to finish. We thank his parents Francesco and Giusy and the sister Beatrice for allowing us to publish the work of "Seba."

References

1. Biley FC. Music as therapy: a brief story. Complement Ther Nurs Midwifery. 1999;5:140–3.
2. Fitch WT. The biology and evolution of music: a comparative perspective. Cognition. 2006;100:173–215.

3. Reisenweay AJ. The development of the flute as a solo instrument from the medieval to the baroque era. Musical Offer. 2011;2(1) https://doi.org/10.15385/jmo2011.2.2.
4. Artaud P-Y, Dale C. Aspects of the flute in the twentieth century. Contemp Music Rev. 1993;8:131–216.
5. Ardal P. The flute. New Haven & London: Yale University Press; 2003.
6. Darwin C. The descent of man, and selection in relation to sex. London: John Murray; 1871. p. 56–7.
7. Pankseep J. Affective neuroscience: the foundations of human and animal emotions. New York: Oxford University Press; 2004.
8. Turk M, Turk I, Otte M. The Neanderthal musical instrument from Divje babe I cave (Slovenia): a critical review of the discussion. Appl Sci. 2020;10(4):1226. https://doi.org/10.3390/app10041226.
9. Landels JD. Music in ancient Greece and Rome. New York: Routledge; 1999.
10. Tarrant RJ, editor. P. Ovidi Nasonis metamorphoses. Oxford: Oxford University Press; 2004.
11. Sachs C. The history of musical instruments. North Chelmsford: Courier Corporation; 2012.
12. Galante E, Lazzari G. Il flauto traverso: storia, tecnica, acustica. Torino: EDT; 2003.
13. Sardelli MF. Vivaldi's music for flute and recorder. London and New York: Routledge Taylor & Francis Group; 2016.
14. Ardal P, Lasocki D. Bach and the flute: the players, the instruments, the music. Early Music. 1995;23:9–29. http://www.jstor.org/stable/3137801.
15. Weaver LR. The Orchestra in early Italian opera. J Am Musicol Soc. 1964;17(1):83–9. https://doi.org/10.2307/830033.
16. Felicioni M. L'evoluzione del flauto traverso- Dal traversiere allo strumento moderno. Raleigh: Lulu Press; 2009.
17. Woo E. Pioneering jazz flutist had a 'peppery' style, Los Angeles Times, June 15; 2013. https://www.latimes.com/archives/la-xpm-2013-jun-14-la-me-sam-most-20130615-story.html. Accessed 21 Dec 2021.
18. Byl NN, McKenzie A, Nagarajan SS. Differences in somatosensory hand organization in a healthy flutist and a flutist with focal hand dystonia: a case report. J Hand Ther. 2000;13(4):302–9.
19. Cossette I, Sliwinski P, Macklem PT. Respiratory parameters during professional flute playing. Respir Physiol. 2000;121(1):33–44.
20. Cossette I, Monaco P, Aliverti A, Macklem PT. Chest wall dynamics and muscle recruitment during professional flute playing. Respir Physiol Neurobiol. 2008;160(2):187–95.
21. Artigues-Cano I, Bird HA. Hypermobility and proprioception in the finger joints of flautists. J Clin Rheumatol. 2014;20(4):203–8.
22. Kleber B, Zeitouni AG, Friberg A, Zatorre RJ. Experience-dependent modulation of feedback integration during singing: role of the right anterior insula. J Neurosci. 2013;33(14):6070–80.
23. Margulis EH, Mlsna LM, Uppunda AK, Parrish TB, Wong PC. Selective neurophysiologic responses to music in instrumentalists with different listening biographies. Hum Brain Mapp. 2009;30(1):267–75.
24. Goad PJ, Keefe DH. Timbre discrimination of musical instruments in a concert hall. Music Percept. 1992;10(1):43–62. https://doi.org/10.2307/40285537.
25. Hadlock H. Sonorous bodies: women and the glass harmonica. J Am Musicol Soc. 2000;53(3):507–42. https://doi.org/10.2307/831937.
26. Gallo DA, Finger S. The power of a musical instrument: Franklin, the Mozarts, Mesmer, and the glass armonica. Hist Psychol. 2000;3(4):326–43.
27. Meyer V, Allen KJ. Benjamin Franklin and the glass armonica. Endeavour. 1988;12:185–8.
28. Ashbrook W. Donizetti. London: Cassell; 1965. p. 417.

Neurological Diseases in Popular Songs

11

Bruno Colombo and Luca Bosco

What Is Popular Music?

Popular music, quoting Serge Denidoff, is like a unicorn: "everyone knows what is supposed to look like, but no one has ever seen it" [1]. Generally speaking, popular music consists of a hybrid of musical influences, styles, and traditions. Admittedly, it is also an economic product, which is invested with ethical, moral, and ideological significance by the majority of its consumers. In English-speaking countries, we find two interpretations of the word popular music: music "of the people" or music that "appeals to many," with reference to its popularity. In English, music of oral tradition is referred to as folk music or traditional music. Thus, popular music can be defined as music with a wide circulation, to be listened to through the media, such as the radio, the television, and social channels. According to the Italian musicologist L. Pestalozza, popular music can be defined as "extra-cultured," thus indicating a music genre outside academic boundaries and in a context not classifiable as "classic" or "cultured" [2]. Actually, popular music must be seen as a proper cultural unit, something that culture has defined as an element, distinct and different from others. The popular music genre or style is therefore a collection of musical facts (activities around any kind of sound events), and its performance is ruled by socially and culturally accepted norms [3]. These norms are of various kinds: formal technical, semiotic, ideological, and behavioral. The peculiar meaning of popular music is the product of a sort of circular process, operating at different cultural levels in social, institutional, and personal domains. Popular music has long been defined as ephemeral, silly, and frivolous, but it has proved to have a lasting cultural impact, worthy of scholarly

B. Colombo (✉) · L. Bosco
Department of Neurology, San Raffaele Hospital, Vita-Salute University, Milano, Italy
e-mail: colombo.bruno@hsr.it

B. Colombo (ed.), *The Musical Neurons*, Neurocultural Health and Wellbeing,
https://doi.org/10.1007/978-3-031-08132-3_11

work (i.e., the International Association for the Study of Popular Music). Studying popular music and the so-called "taste communities" linked with it can bring us into specific aspects of popular culture. Communities are to be understood as fluid but made of people linked by a sort of continuous act of imagination. Consequently, cultural units, as are, for example, musical genres and here specifically popular music, can be understood as complex cultural signs. They are labels for a new reality derived from the growth and recognition of a newfound segment within popular culture. For this reason, popular music can help us to better understand a particular way of life, whether of a people, an historical period, or a community. Furthermore, it can analyze the works and practices of intellectual and especially artistic, spiritual, and aesthetic activity. Popular music can be used as a sort of sharp lens that can help us to better comprehend the ways that groups and communities think of themselves culturally and politically.

Hence, if popular music is the expression of a popular culture, song lyrics can be analyzed as ways compiled by artists to indicate a cultural position relative to a specific problem. A song is not only the telling of a story, but it is also in its self-part of the story and is intertwined with the cultural and social times and developments it produces and receives, within the different media by which it exists.

To this aim, and this is the objective of this chapter, appreciating how neurological disorders are experienced and/or interpreted within the framework of popular music can help us understand the experience of certain issues such as dementia, epilepsy, and migraine within the mainstream. Furthermore, it can enable us to assess how pathologies are included in artistic narratives and which insights have then emerged as reflections of the popular music of a certain period. Lyrics of popular songs have the special characteristic of being able to reflect the mainstream way of thinking, providing a unique and accessible point of view onto the values and contemporary life experiences of a particular people and/or generation, quite often directly and without embellishment. Moreover, they provide an insight into what is permitted in current music publication standards, as it is the case for discriminatory or obscene lyrics.

Online music sites were searched for English popular songs with the selected word in the text or title. We then analyzed the most significant results according to music genre, artist demographics, and general approach to the theme.

Due to the variable degree of representation of these disorders in the lyrics of popular English music during the past 30 years, we pursued an intriguing journey into the ever-changing habits and morals of modern western society, listening and analyzing songs that were heard, danced, and sang by roughly three generations of people worldwide.

11.1 Migraine

Take away the sensation inside
Bitter sweet migraine in my head
It's like a throbbing tooth ache of the mind
I can't take this feeling anymore
Give me novocaine—Green Day (2004)

Migraine is a common denominator of human experience, likely to be encountered at either first or second hand by many within population. Migraine has a particular impact on young people, a group that is prone to influence by media and popular culture. So is not surprising that artists have on occasion used such conditions as source materials for elaboration in their narratives. In this study we aimed at reaching a better understanding of the ways in which migraine is portrayed in popular music, comparing song texts released in the last 30 years.

The word "migraine" has been found to be present in 76 songs. The artists who have used these words are all males, with a few exceptions. All the musicians have less than 40 years. During the three decades the prevalent genre saw an increase of indie artists, with the most prevalent genre remaining hip-hop, rap, and, recently, trap. In all cases, migraine is represented with a negative connotation, associated with frustration, despair, anger, and solitude. Migraine is described as an intrusive illness, a social disease, a difficult to cure physical and emotional pain.

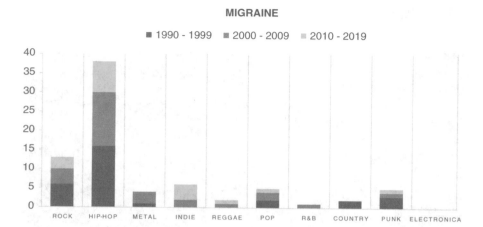

MIGRAINE

■ 1990 - 1999   ■ 2000 - 2009   ■ 2010 - 2019

Box 11.1: Migraine: An Example from another Time, G. Rossini

Gioachino Rossini (1792–1868) was an Italian composer of fervid and brilliant creativity. His life was one of alternating successes (he wrote both comic and dramatic operas such as *L'Italiana in Algeri, Il barbiere di Siviglia*, and *Guglielmo Tell*) and retreats due to his rather bad health. Since he was in his forties Rossini in fact suffered from a debilitating chronic gonorrhea. During the last years of his life, he was also affected by chronic bronchitis, worsened by his considerable obesity. He then also suffered the consequences of a cerebral stroke. Rossini died from infectious complications of a rectal tumor, operated by the French surgeon Auguste Nélaton in November 1868. His very poor state of health also influenced his mood and in times of deep depression he complained of headaches and suicidal ideations.

This is how Rossini playfully describes migraine in his opera *L'Italiana in Algeri*. This is a lighter, less gloomy depiction compared to those found in current popular music but manages to show very well a precise and specific symptomatic picture.

TADDEO, MUSTAFÀ, ELVIRA, ISABELLA, ZULMA, LINDORO, HALY Va sossopra il mio cervello, Sbalordito in tanti imbrogli; Qual vascel fra l'onde e i scogli Io sto/Ei sta presso a naufragar.	TADDEO, MUSTAFÀ, ELVIRA, ISABELLA, ZULMA, LINDORO, HALY My brain is in uproar Stunned by so many tangles; Like a vessel between the waves and the rocks I am/he is about to be shipwrecked.
CORO Va sossopra il suo cervello; Ei sta presso a naufragar.	CHOIR His brain is in uproar; He is about to be shipwrecked.
ELVIRA Nella testa ho un campanello Che suonando fa din din.	ELVIRA In the head I have a bell Which goes din din.
ISABELLA e ZULMA La mia testa è un campanello Che suonando fa din din.	ISABELLA and ZULMA My head is a bell Which goes din din.

LINDORO e HALY Nella testa ho un gran Martello Mi percuote e fa tac tà.	LINDORO and HALY In the head I have a great hammer Which hitting goes tac tà.
TADDEO Sono come una cornacchia Che spennata fa crà crà	TADDEO I'm like a crow What plucked goes crà crà
MUSTAFÀ Come scoppio di cannone La mia testa fa bum bum.	MUSTAFA Like a cannon blast My head goes bum bum.

11.2 Epilepsy and Seizure

*You're f***ing with me, nigga you f** around
and catch a seizure or a heart attack
You better back the f** up before you get
smacked the f** up*
Hit 'Em Up—2Pac (1996)

Throughout history, epilepsy has been the object of stigma and marginalization. Many of the historical associations of epilepsy are with horror, madness, and lunacy and can still be found in most of popular music lyrics. Its unique visual impact in the witnesses of its manifestations, along with the usual initial presentation during childhood or adolescence, made this illness permeate in modern music, especially among young artists.

The word "seizure" has been found to be present in 104 songs, with more than 60% belonging to hip-hop music, which remained the most prevalent genre across all the three decades, with an increasing presence of the trap subgenre during the 2010s. "Epilepsy" has been found in 11 songs, eight of which belonging to the hip-hop genre. The main theme associated with seizure (or epilepsy) in these lyrics is by far violence. Most hip-hop artists associate seizures with the outcomes of a beating or a traumatic brain injury, usually portraying a concrete visual description of the scene. This can be easily linked to the "street-combat" theme frequently adopted in the lyrics of the genre, with depictions of violent gang actions usually culminating with the physical overwhelming of rivals.

On the other hand, the most extreme subgenres of heavy metal presented a tendency to elaborate epilepsy more metaphorically, as a curse, or a demon, associating it with madness and gore acts.

In some other cases, epilepsy is presented as a state of sexual ecstasy and dance euphoria, or as a metaphor to describe the trance of frenetic dance moves, withholding its general negative connotation, but this is usually an exception.

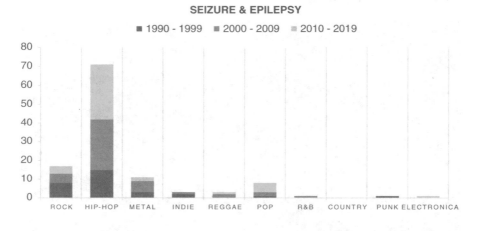

11.3   Insomnia

Getting stressed, making excess mess in darkness
  No electricity, something's all over me, greasy
  Insomnia, please release me
  And let me dream of making mad love to my girl on the heath
  Tearing off tights with my teeth
  But there's no release, no peace
  I toss and turn without cease
  Like a curse, open my eyes and rise like yeast
  Insomnia—Faithless (1995)

Sleep, dreams, and the more specific theme of insomnia are ubiquitous in pop culture, and music constitutes no exception, both today and in the past. Sleep has been a popular subject in stories, regardless of the form, for centuries. And it is also clear, that for centuries, artists have been committing their mesmerizing nocturnal thoughts, illogical musings, or flashes of creative genius to artistic production.

The word "insomnia" has been found in 102 songs, equally distributed across rock, hip-hop, metal, and, to a lesser extent, pop, with a relative increase of the hip-hop presence along the years. Common themes often associated with insomnia are grief and psychiatric disorders. Emotions most frequently associated with insomnia are unsettling general distress, sadness, and anxiety.

Insomnia is usually described in a negative fashion, as an illness capable of deteriorating the mental health of the artist. A shared metaphor across songs is that of an intrusive but acquainted entity, with which the artist meets every night, and that prevents him from getting the peace of mind, in sleep. Some exceptions to this portrayal see the lack of sleep as an opportunity able to increase the potential of artistic production.

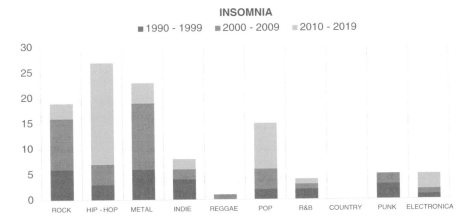

11.4   Dementia

Here for you and all mankind
I've lost my mind
Psychotic rounds in rabid dementia
I won't be fine
Welcome to the Family—Avenged Sevenfold (2010)

Dementia represents a broad category to define brain diseases that cause a long-term and gradual decrease in the ability to think and remember in a way to affect a person's daily functioning.

Due to its high prevalence in the general population and the overwhelming burden of care onto families, societies, and governments, it is expected that some of all this struggle is going to be represented in popular song lyrics.

The word "dementia" has been found to be present in 55 songs, with more than half of those belonging to heavy metal subgenres, predominantly death and goth metal.

However, if dementia was almost exclusively dealt by these genres in the past (with 10 out of 12 songs belonging to heavy-metal extreme subgenres in the 1990–1999 decade), in the last years we observed a tendency toward the emerging of the theme in other genres as well, such as rock, indie, hip-hop, and pop.

Themes usually associated with the word dementia are hopelessness, solitude, and death. Dementia is invariably described under a negative light, with old age and symptoms of mental illnesses being regularly co-depicted. The stereotypical image is that of a person who loses his memory and identity, behaves unpredictably, and suffers. The predominant emotions associated with dementia in those lyrics are fear, loneliness, and despair.

Also due to the gruesome themes usually dealt by the predominant genres, we found dementia to be by far the neurological condition carrying the most stigma, and being worse perceived, in popular music.

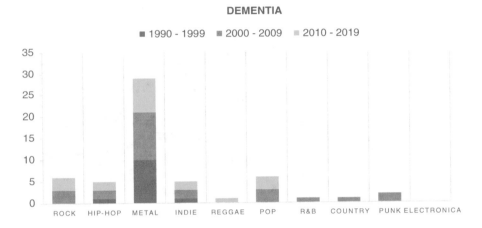

11.4.1 Conclusions

Since the lyrics of pop songs reflect the mainstream way of thinking of young generations, neurological diseases still have a firmly irremovable stigma of chronic and hopeless disease. During the last three decades, little signs of improvements in the social background of neurological diseases have transpired in popular songs.

11.4.2 Other Neurological Diseases in Music

Other neurological diseases have been variably cited or adopted as themes in popular song and music in general. Due to the low frequency of appearance of most of these disorders in lyrics of popular songs, we were not able to include them in our descriptive analysis. On the other hand, we felt the need to report at least the two examples below that we consider particularly worthy of mention in the way the artist can differently deal, and in some cases cope, with this vast and complex class of diseases.

The first of these two examples consists of our English translation of the open letter from the Italian songwriter and composer Bruno Lauzi (1961–2006), furiously dedicated to his relentlessly progressive neurological disorder, embodied by Mr. Parkinson. In this letter, Lauzi keeps intact his extraordinary verve, frankness, and great sense of irony, while also promoting initiatives to raise funds for the study and assistance of Parkinson's patients.

Open Letter to Mr. Parkinson: Bruno Lauzi (2006)

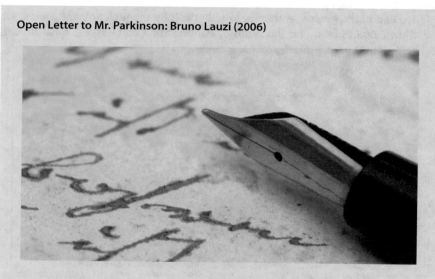

Dear Sir,

It is not with pleasure that I am writing you this letter, but on the other hand I should have talked to you face to face, and face you in person, and endure your sneaky way of doing which is the worst thing to make lose patience even to a saint, let alone me.

I am writing to you, as you can see, with the computer, because my handwriting has become illegible and so tiny that my collaborators have to use a magnifying glass to be able to decipher it.

Why am I writing this to you?

It is easy to say: I have overcomed with a certain ease the embarrassment that you (and I write that without a capital letter on purpose, as you do not deserve it) created by publicly asking for my hand, and obviously obtaining it.

Living with a retired English officer, already condemned in Punjab for repeated attempts of neurological violence on beings of any kind (certain things tend to come out, as you can see ...) was not easy, as mine is an old-fashioned family, and did not appreciate.

BUT NOW YOU ARE EXAGGERATING, sir, I have to tell you. When it is too much, it is too much, and too much even of a good thing! There is an Arab proverb that says: "If you have a friend made of honey, don't lick it all", BUT YOU TAKE ADVANTAGE OF EVERY RELAXATION, OF THE LOWERING

OF THE GUARD IN THE EVERYDAY BATTLE, you forbid us to think about anything else, counting on the superficiality with which I faced the onset of your evil ... you know, artists are unconscious butterflies ... no, old goat, it will not be easy, neither with me nor with the others, the Resistance has begun. Because, you see, my sick brothers and sisters and I have so many things to do, a life to carry on, much better than that!

From now on I promise that I will be more attentive to the advice of my doctors, and that I will do more to help them raise the necessary funds for research. Indeed, on the theme of solidarity I'm willing to bet my hand, that hand that, painted and silk-screened, is acting as a pedestal for a poem against you, colonel of my boots, and functioning as an incentive to give ... yes, because for any offer to the research it will be sent "THE HAND" as a souvenir and memento...

We are many, and many hands will rise against you and will try to return you blow for blow until they will be able to grab you by the scruff and send you to Hell, where you belong, filthy beast, dung of the devil, cross without delights ... You have my word, the one of this little man, for many a little funny man, for others a little pathetic man, but still a man that is living the dream of being able, one day, not far away, to slap you.

With a steady hand.

I wish you the worst, and hope to see you never again.

11.4.3 Arie Perry's Neuropathology Songs

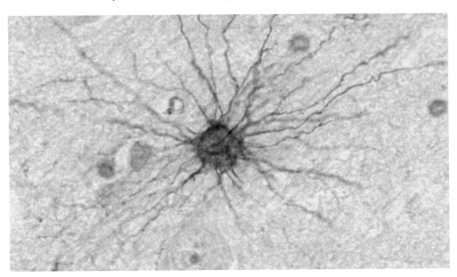

Tenor Arie Perry's neuropathology songs arise from a completely different need as compared to what we have dealt with until this point. Professor Perry, neuropathologist at UC San Francisco, experimentally wrote songs to be used as tools to help medical students grasp the basics of neuropathology in different diseases, such as multiple sclerosis, brain tumors, and neurodegenerative disorders, similarly to what is routinely done with pre-school children in the learning of the fundamentals. While mastering an extensive knowledge of neurobiology is surely more complex than just memorizing the alphabet song, putting some of the most essential facts in music has proven to be a helpful and surely singular way to approach the learning of complex and often intimidating topics.

> **Multiple Sclerosis—Arie Perry (2010)**
> *Optic neuritis, spinal pathology,*
>   *periventricular plaques,*
>   *Dawson's fingers fan out radially, surrounding vascular tracks*
>   *They're sharply demarcated, I don't know why;*
>   *looks gray and sunken to the naked eye*
>   *Cortex may be involved, this we know now,*
>   *cognitive loss occurs, perhaps this is how, oh MS!*

References

1. Jones G, Rahn J. Definitions of popular music: recycled. J Aesthet Edu. 1977;11(4):79–92.
2. Pestalozza L. L'opposizione musicale. Milano: Feltrinelli; 1991.
3. Fabbri F. Il suono in cui viviamo. Milano: Feltrinelli; 1996.

Conclusion

12

Bruno Colombo

Art and music are a representation of what is and isn't visible.

But they are also the mental prefiguration of what does not yet exist outside the mind of the artist or composer.

Art and music are thus alive in the imagination and are, to this effect, close to philosophy and science.

There is no science and philosophy devoid of imagination, and there is no art and music without the involvement of creating thinking and research.

These three worlds, art, philosophy, and science, all primarily aim at the pursuit of truth and beauty.

Elements of suggestion, enchantment, and reflection invite the artist, the scientist, and the philosopher to achieve a sense, an image, and a piece of music.

Significance is a challenge in time, a need to leave a legacy and a perpetual overcoming of limits and paradigms.

And all this is contained in our neural connections, which are nonreplicable, bold, and unique.

Let's teach them to mature and grow, for our benefit and for that of our society.

We are artists, philosophers, and scientists. But above all, we are humans, responsible for our future.

B. Colombo (✉)
Department of Neurology, Headache Clinical and Research Center, Ospedale San Raffaele, Vita-Salute University, Milano, Italy
e-mail: colombo.bruno@hsr.it

© The Author(s), under exclusive license to Springer Nature
Switzerland AG 2022
B. Colombo (ed.), *The Musical Neurons*, Neurocultural Health and Wellbeing,
https://doi.org/10.1007/978-3-031-08132-3_12

"applauses" courtesy of Lorenzo Colombo

Printed in the United States
by Baker & Taylor Publisher Services